STAND-ALONE SOLAR ENERGY

PLANNING, SIZING AND INSTALLATION OF

STAND-ALONE PHOTOVOLTAIC SYSTEMS

Oliver Style

First English edition

August 2013

IT CA

Appropriate Technology

STAND-ALONE SOLAR ENERGY

PLANNING, SIZING AND INSTALLATION OF STAND-ALONE PHOTOVOLTAIC SYSTEMS

First English edition
August 2013

ISBN: 978-84-616-5811-4

Graphic design: Richard Grove
Translation: Jennifer Stanley Smith, Oliver Style

ITACA
Appropriate Technology

This book is dedicated to KIPTIK, CATAS, and the autonomous communities of Chiapas, Mexico.

My thanks to the Concern America team, Richard Grove, Santiago Arnalich, and Jennifer Stanley Smith.

This publication has been made possible thanks to the support of the NGO Concern America.

INDEX

ABBREVIATIONS AND SYMBOLS

≈	Approximately
≤	Less than or equal to
≥	Greater than or equal to
A	Amp
AC	Alternating Current
AGM	Absorbed Glass Matt
DC	Direct Current
DoD	Depth of Discharge
PSH	Peak sun hours
kW	Kilowatt
kWh	Kilowatt-hour
kWh/m2	Kilowatt-hour per metre squared
Lm	Lumen
MPPT	Maximum Power Point Tracking
NOCT	Normal Operating Cell Temperature
SPV	Stand-alone photovoltaic system
SoC	State of Charge
SOC	Standard Operating Conditions
STC	Standard Test Conditions
V	Volt
W	Watt
Ω	Ohm

1 INTRODUCTION

1.1 ABOUT THIS BOOK

This book is a basic guide for technicians, installers and operators who want work with stand-alone photovoltaic systems. It contains everything you need to get results and complete a successful project, with illustrations and worked examples to make the learning process easier.

> ➢ Chapter 1 gives a general overview of solar photovoltaic energy and the key concepts for working with electricity.

> ➢ Chapter 2 covers the solar resource and how it can be harnessed in an SPV system.

> ➢ Chapters 3 to 9 describe the main components of a system, enabling the reader to understand their function and to select the most appropriate equipment.

> ➢ Chapters 10 and 11 present the steps for the sizing and economic analysis of a low power stand-alone systems.

> ➢ Chapters 12 and 13 deal with topics related to installation, commissioning, operation and maintenance.

> ➢ Chapter 14 contains a Glossary of Terms.

> ➢ Chapters 15 and 16 include a Bibliography and Appendices with additional technical information.

1.2   SOLAR PHOTOVOLTAIC ENERGY

A stand-alone or off-grid photovoltaic system (SPV) converts energy from the sun into electrical energy, storing it in a battery to be used later on. It is a system that does not require a connection to the electrical grid, as it functions independently to provide power for appliances and lighting. These kinds of systems are well suited to remote places, where there is no connection to the grid, where energy consumption is low and where there is a good solar resource. In certain places they can be the most suitable solution for providing electricity to a building or home.

With the prices of photovoltaic modules constantly falling and the imminent improvement in storage technologies, SPV systems will become an ever more accessible option for the 1,400 million people in the world who still live without electricity. SPV systems can supply energy to clinics, hospitals, schools, communication posts, homes and water pumping systems, among other applications.

Figure 1: Mounting the panels for an SPV system

ADVANTAGE	DISADVANTAGE
The sun is a free and renewable source of energy, available in most sites.	Not appropriate for large installations with a high energy demand.
There are no recurring fuel costs.	The batteries must be replaced periodically.
Operation and maintenance are relatively simple.	The supply of energy depends on the amount of solar radiation available.
They are well suited to remote places and can be more economical than grid connections.	Sometimes there is no local specialised technical service for the repair of equipment.
They can be enlarged at a later date.	The initial cost is high.
They do not emit noxious gases or make any noise while in operation.	End users need basic training in the operation and maintenance of the system.

Table 1: Advantages and disadvantages of an SPV System

Understanding the limits of any technology is important. Some of the advantages and disadvantages can be found above in Table 1.

For applications where there is a higher energy demand, it is common to combine a stand-alone system with another source of energy, such as an electrical generator running on diesel or petrol, or a wind turbine. This type of system is called a hybrid system, and has the advantage of reducing the initial cost of investment and ensuring a constant energy supply.

1.3   A STAND-ALONE PHOTOVOLTAIC SYSTEM

In its most basic form, a stand-alone photovoltaic system (SPV) consists of:

> **Array:** the photovoltaic modules that converts solar radiation to electrical energy.

> **Battery:** the batteries where the energy from the module is stored.

> **Charge controller:** a device which controls the energy produced by the module, the level of battery charge, and the energy consumed by the equipment.

> **Load:** energy-consuming appliances (lamps, radios, computers, televisions etc.).

Finally, we have the cables (which transport the energy between the different components of the system), circuit protection devices (fuses and/or circuit breakers) and switches.

1.4 AN SPV SYSTEM IN DIRECT CURRENT

Energy generated by photovoltaic modules and stored in the batteries is direct current (DC) electricity. A DC circuit is characterized by the flow of energy in only one direction, with conductors that have a negative (-) or positive (+) polarity.

Figure 2 shows a simplified diagram of a direct current SPV system. This type of system generally works at a very low voltage (12V or 24V DC), and is primarily used for lighting and for DC equipment with a low energy consumption.

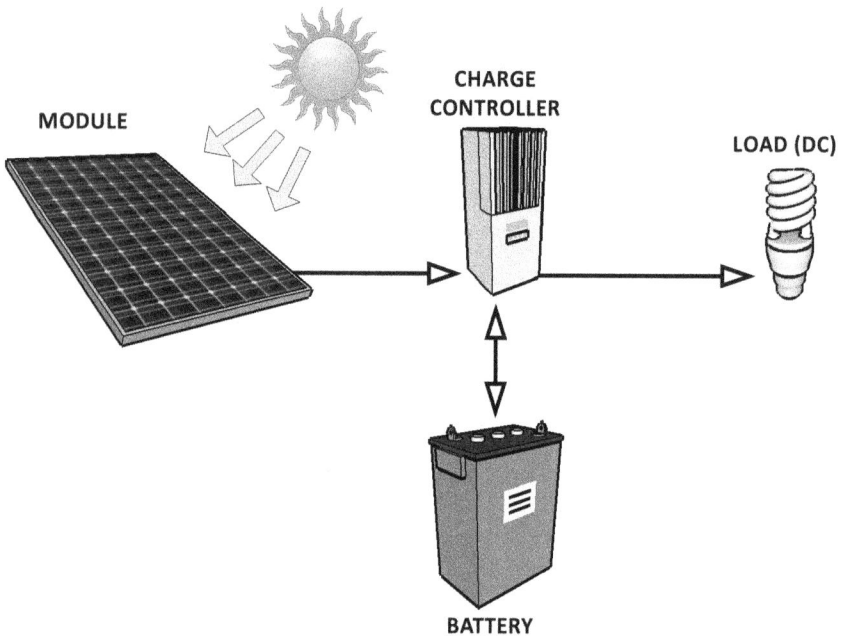

Figure 2: An SPV system in Direct Current (DC)

1.5 AN SPV SYSTEM IN DIRECT AND ALTERNATING CURRENT

Alternating current (AC) electricity changes direction periodically and it is the kind of electricity supplied by national electricity grids. The vast majority of appliances available on the market consume AC electricity. In order to use these appliances with an SPV system, you need a device which converts DC to AC: an inverter. Figure 3 shows a simplified diagram of an SPV system, supplying DC to a lamp and AC to a hi-fi system.

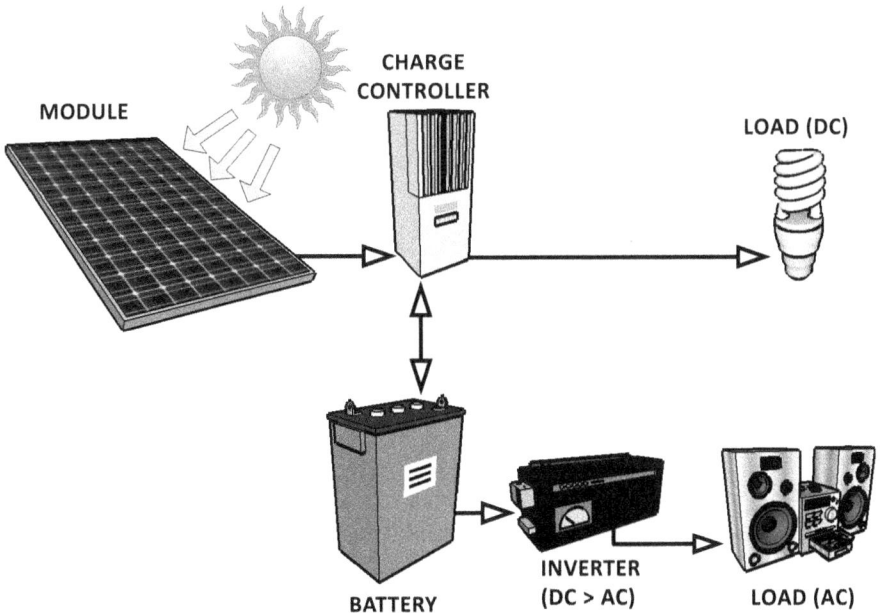

Figure 3: SPV system with AC and DC

1.6 PLANNING

SPV systems can be an efficient and economical way of providing renewable energy for many years. However, to ensure that the system has a long useful life, it's important that the design and installation of the system are carried out correctly. Also, end-users or system operators have to take responsibility for the operation and maintenance of the system to ensure that it continues to function well. In Guatemala in 1999, the NGO Fundación Solar (Solar Foundation) carried out an evaluation of 124 stand-alone systems installed in the Zacapa area. The results of the evaluation showed that 45% of the systems did not work. At a global level, 1 in 4 SPV systems fail. Oops...

This book will help you with the technical aspects of SPV systems to ensure that your system has the best chance of working smoothly year after year. Before starting, keep the following points in mind, which are really nothing more than using a bit of common sense:

➢ **Simplicity:** try to find the simplest technological solution to achieve your objective.

➢ **Planning:** dedicate enough time to the planning of the project and it'll have more chance of success.

➢ **False economies:** the available budget is almost always a limiting factor. However, if you specify the cheapest equipment for a system to save money today, it will generally be low quality and will cost you more tomorrow. Weigh up the economic criteria in the short, medium and long term.

> ➤ **Training:** try to ensure that the system operators or the end-users receive adequate training in the operation and maintenance of the system. The successful application of any technology depends, ultimately, on the user.

> ➤ **If in doubt, ask:** if you're not familiar with electricity or PV systems, don't keep your doubts to yourself. Consult an expert in your area and get the support you need.

> ➤ **Read the damn manual:** often a few hours of trial and error will save you minutes of reading the manual (!). Make your life easier, read the manuals and the technical data sheets before buying, installing or using any equipment!

Figure 4 shows a flow diagram for the planning of an SPV system.

Codes and regulations regarding SPV systems and electrical installations vary from country to country. Consult your national codes and make sure you comply with the requirements.

1.7 KEY CONCEPTS

To size, install and maintain an SPV system, it is necessary to understand some of the basic concepts of electrical energy. Table 2 contains a summary of the most common terms and their definitions.

TERM	SYMBOL	UNIT	DEFINITION
Power	W, kW	Watts, Kilowatts	The quantity of energy delivered
Energy	Wh, kWh	Watt-hours, kilowatt-hours	The capacity to do work Energy = power × time
Current	I	Amps (A)	The intensity of the flow of electrons through a circuit
Direct Current	DC	Amps (A)	The flow of an electrical charge which does not change direction
Alternating Current	AC	Amps (A)	The flow of an electrical charge which changes direction periodically
Voltage	V	Volts (V)	The difference in potential energy between two points of a conductor
Resistance	R	Ohms (Ω)	The property of a conductor which opposes the passage of the current

Table 2: Summary of basic electrical concepts

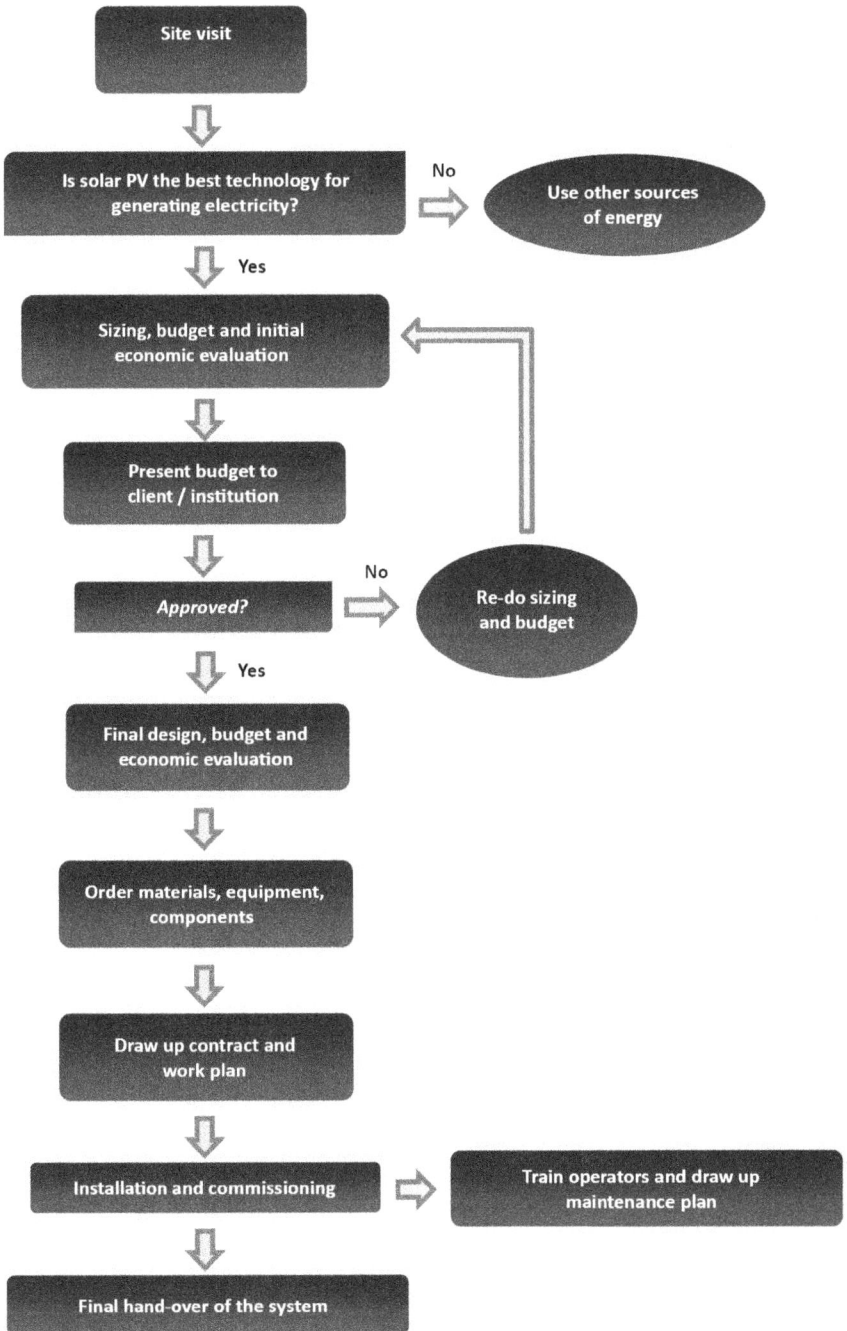

Figure 4: Flow diagram for the implementation of an SPV system

The difference between power and energy is a common point of confusion if you're just starting out with PV systems and electricity. Power is the quantity of energy delivered at a given point, while energy is that power *used for a given length of time.*

Worked Example 1: Power and energy

I have a lamp with a power of 20 watts (20 W). If I turn it on for 1 1/2 hour, how much energy have I used?

Energy (Wh) = Power (W) × Time (h)

Energy (Wh) = 20 (W) × 1.5 (h) = 30 Wh

If you are not familiar with electricity, find someone who is qualified before starting the planning, design or installation of a system.

1.8 THE POWER EQUATION

There are two laws that will help you work with SPV systems: the Power Equation and Ohm's Law. The Power Equation describes the relationship between Power, Current and Voltage:

THE POWER EQUATION
Power (P)   =   Current (I)   ×   Voltage (V)

If we know the Current (I) and Voltage (V) of a circuit, we can work out the Power (W):

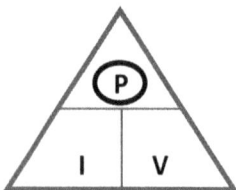

Power (P) = Current (I) × Voltage (V)

If we know the Power and the Voltage, we can calculate the Current:

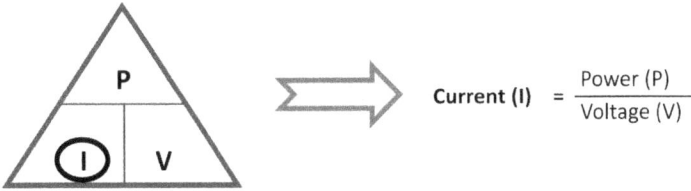

$$\text{Current (I)} = \frac{\text{Power (P)}}{\text{Voltage (V)}}$$

If we know the Power and the Current, we can calculate the Voltage:

$$\text{Voltage (V)} = \frac{\text{Power (P)}}{\text{Current (I)}}$$

Worked Example 2: Power

I have a lamp connected to a 12 **volt** battery which consumes 2 amps (**current**). What is its **power** consumption in watts?

Power = 12 V × 2 A = **24 W**

Worked Example 3: Current

I have a television with a power of 110W connected to an inverter with an output voltage of 220 V. How much current does the television consume?

$$\text{Current} = \frac{110\text{ W}}{220\text{ V}} = \textbf{0.5 A}$$

1.9   OHM'S LAW

Ohm's Law describes the relationship between Voltage, Current and Resistance:

OHM'S LAW (Ω)				
Voltage (V)	=	Current (I)	×	Resistance (R)

If we know the Current (I) and Resistance (R) of a circuit, we can work out the Voltage (V):

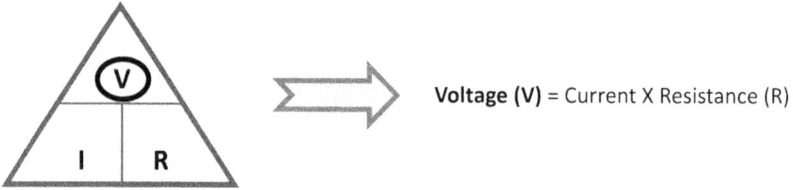

Voltage (V) = Current X Resistance (R)

If we know the Voltage and Resistance, we can calculate the Current:

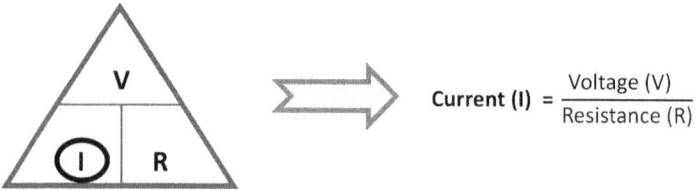

Current (I) $= \dfrac{\text{Voltage (V)}}{\text{Resistance (R)}}$

If we know the Voltage and Current, we can calculate the Resistance:

Resistance (R) $= \dfrac{\text{Voltage (V)}}{\text{Current (I)}}$

1.10 UNITS

When working with SPV systems, various units are used to measure and calculate the demand and production of energy in the system. Table 3 shows the most frequently used units and their conversions.

POWER	
1 watt × 1000 = 1 kilowatt	1 W × 1000 = 1 kW
1 kilowatt × 1000 = 1 megawatt	1 kW × 1000 = 1 MW
ENERGY	
1 watt-hora × 1000 = 1 kilowatt-hour	1 Wh × 1000 = 1 kWh
1 kilowatt-hour × 1000 = 1 megawatt-hour	1 kWh × 1000 = 1 MWh

Table 3: Units and conversions

Worked Example 4: Resistance

I have a circuit with a lamp that consumes a **current** of 2A, connected to a battery with a voltage of 12 V. What is the **resistance** of the lamp in Ohms (Ω)?

$$\text{Resistance} = \frac{12\ V}{2\ A} = 6\ \Omega$$

Worked Example 5: Converting units

I want to connect 4 houses to an SPV system. Two houses have a daily consumption of 500Wh, and the other two each consume 800Wh. In total, how much energy do the houses consume in KWh?

The total consumption of the houses is: (2 × 500 Wh) + (2 × 800 Wh) = 2 600 Wh

$$\text{Total consumption} = \frac{2\ 600\ Wh}{1\ 000} = 2.6\ kWh$$

2 THE SUN

For now at least, the sun's energy is free, and unless something changes very drastically, the sun rises every morning and sets every night. Every year the energy that reaches us this way is approximately 5,000 times more than what we consume worldwide. The sun is still a largely wasted resource that we can use, among many other things, to generate electricity.

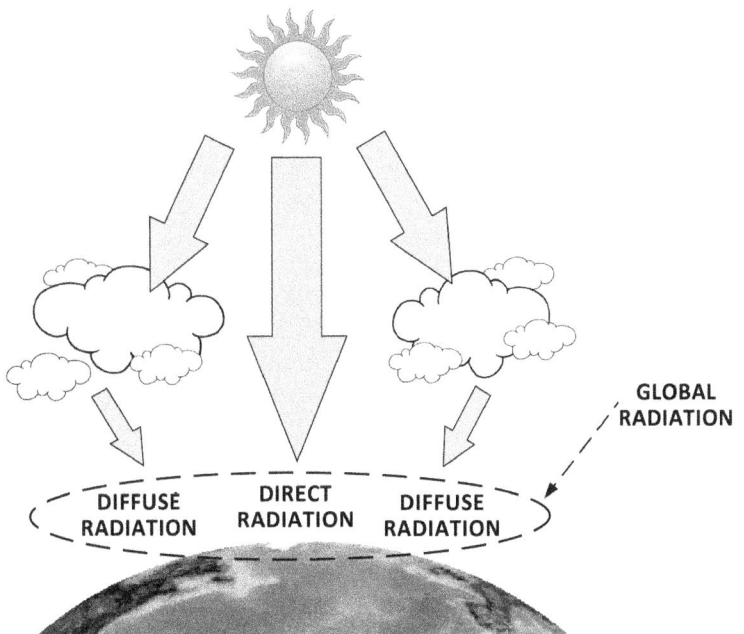

Figure 5: Direct, diffuse and global solar radiation

2.1 SOLAR RADIATION

The energy that reaches us from the sun is referred to as solar radiation, which consists of direct radiation and diffuse radiation. Direct radiation comes in a straight line from the sun. Diffuse radiation reaches us after being reflected off clouds, smog, or dirt. The combination of direct and diffuse radiation is called global radiation (Figure 5).

There are four factors which can affect the amount of solar radiation available in a given place:

➤ **Latitude:** position north or south of the equator.

➤ **Cloudiness:** on a cloudy day, the amount of diffuse solar radiation can be one tenth of what it would be if direct radiation was getting through.

➤ **Humidity:** humidity in the air absorbs solar radiation.

➤ **Atmospheric pollution:** clouds, smog and dust block solar radiation from getting through.

2.2 SOLAR IRRADIANCE

Solar irradiance is a measure of the solar radiation that falls on a given area, and is measured in W/m2, or kW/m^2. When solar energy reaches the atmosphere, it has a power of approximately 1350W/m^2. However, as it passes through the atmosphere it loses power, and when it reaches the surface of the earth it has a maximum power of approximately 1000W/m^2. Figure 6 shows solar irradiance on a horizontal surface over a day in June in Mexico City.

Solar irradiance Mexico City: June

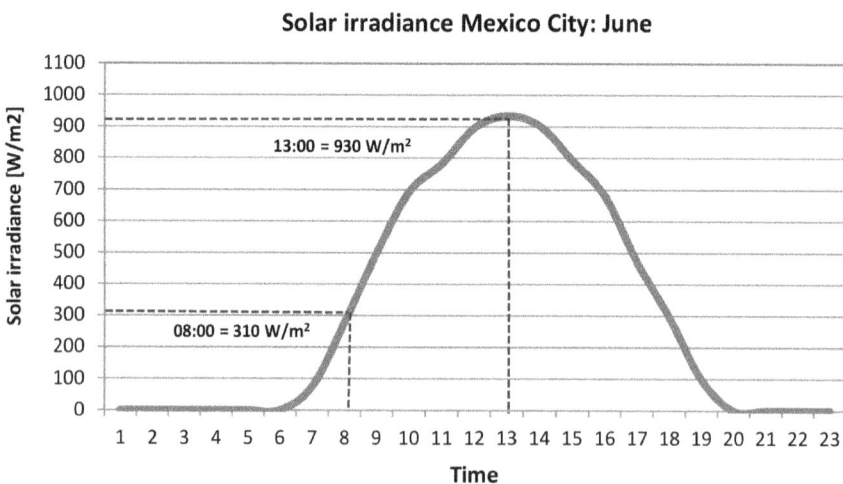

Figure 6: Solar irradiance over a day in June in Mexico City

The vertical axis indicates solar irradiance in W/m². The horizontal axis indicates the time of day. The curve demonstrates how the power of solar irradiance varies at different times of day. At 13.00, when the sun is at the highest point of its daily trajectory, the irradiance is about 930 W/m². However, at 08.00, the irradiance is approximately 310 W/m². The irradiance is weaker in the morning and the evening, on account of two factors: :

a)   the sun's rays have to travel further through the atmosphere to reach the earth and they lose power along the way (Figure 7: the X distances are greater than the distance Y).

b)   a horizontal surface captures less energy from the sun when the sun is lower in the sky (Section 2.1).

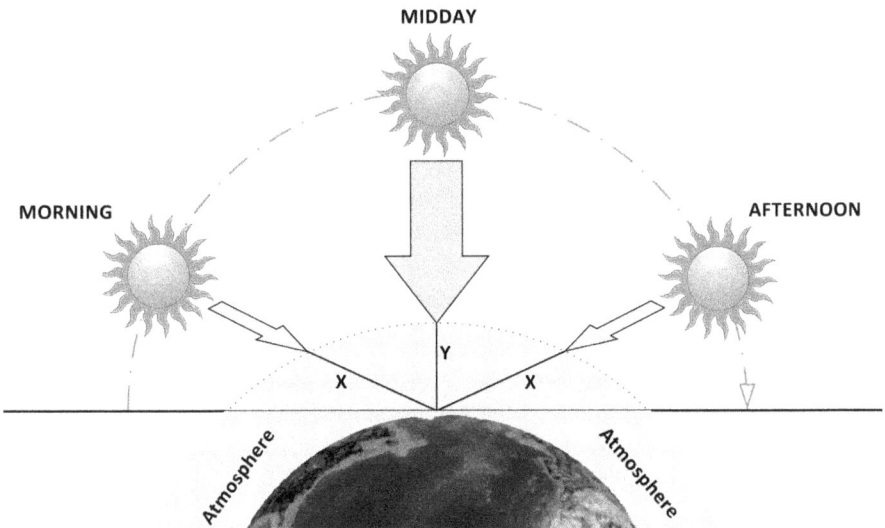

Figure 7: Variation in solar irradiance on a horizontal module over the course of the day

2.3   SOLAR IRRADIATION

Solar irradiation is a measure of the amount of solar energy that strikes a surface over a given period of time. The most frequently used units are Wh/m² per day, or kW/m² per day. For low power SPV systems, solar irradiation is usually referred to as Peak Sun Hours (PSH), equivalent to the hours of the day when the irradiation is 1000 W/m². Figure 8 shows solar irradiation for a day in June when the daily solar irradiation is 7.4kWh/m², or 7.4 Peak Sun Hours.

To size an SPV system, you need to know the solar irradiation or Peak Sun Hours of your location for every month of the year (Appendix 16.5 contains solar irradiation figures for a large number of Latin American countries).

Solar irradiance Mexico City: June

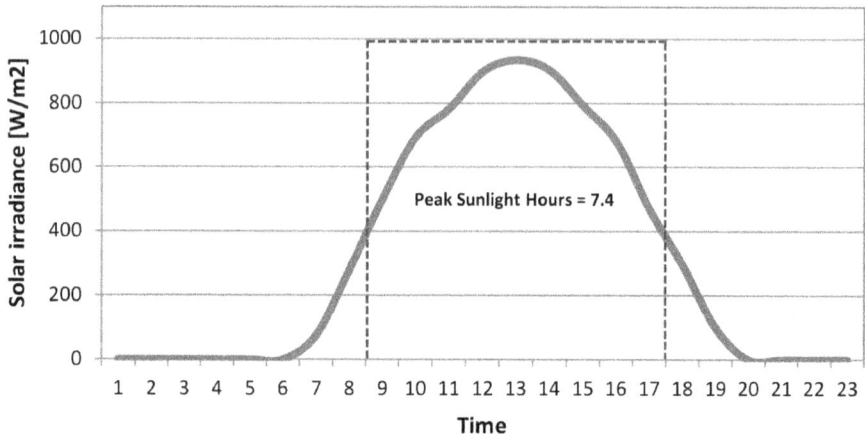

Figure 8: Peak Sun Hours in Mexico City

2.4 ANGLE OF INCIDENCE

The angle at which a ray of sun hits a surface is called the angle of solar incidence. The closer it is to 90º, the greater the amount of energy received. Figure 9 shows the angle of incidence on a horizontal photovoltaic module at midday (90º) and in the evening (33 º). The amount of solar irradiation that the module receives in a horizontal position is greater at midday than in the morning or the evening, because at midday the angle of incidence is 90º.

In order to maximize the collection of solar energy, ideally the modules would rotate to maintain the angle of solar incidence at 90º throughout the day. This is how sunflowers manage to capture so much solar energy, because they rotate to follow the sun and try to keep the angle of incidence at 90º (See Figure 10).

This phenomenon can be seen in larger solar PV systems which have what is called a tracker system, where a small electric motor rotates the modules to follow the sun. This increases the amount of energy the generator captures by up to 30%. For less powerful SPV systems (≤ 500 Wp), automatic trackers are not suitable, as they add unnecessary expense and maintenance requirements. Figure 11 shows a tracker for a grid-connected photovoltaic system.

There are less complicated tracker systems, which may have one or two axles and can be manual or automatic. These increase the yield by a lesser amount but make maintenance simpler (Figure 12).

MIDDAY

MORNING

ATFERNOON

90°

33°

HORIZONTAL PHOTOVOLTAIC MODULE

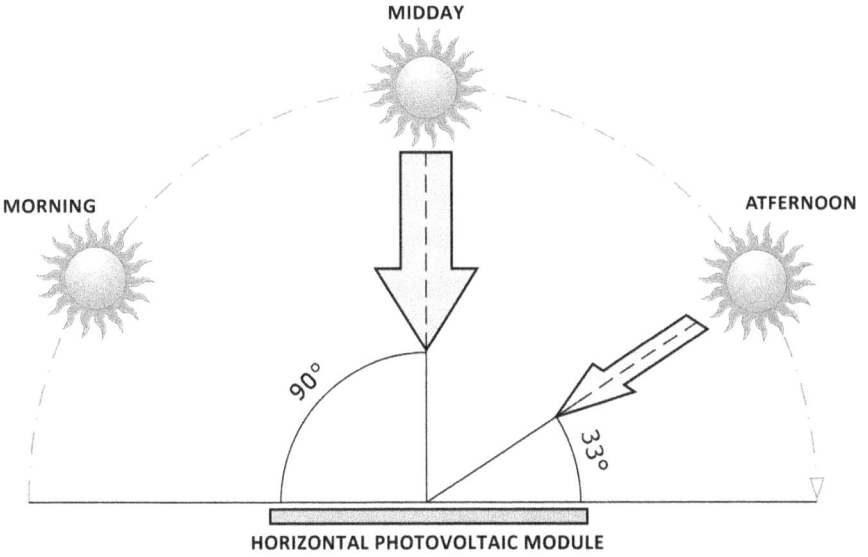

Figure 9: Angle of incidence on a horizontal module

Figure 10: Sunflowers with their inbuilt tracker system

Figure 11: Solar tracker system for a grid-connected array [Source: Premier Power]

Figure 12: Tracker systems with one or two axles

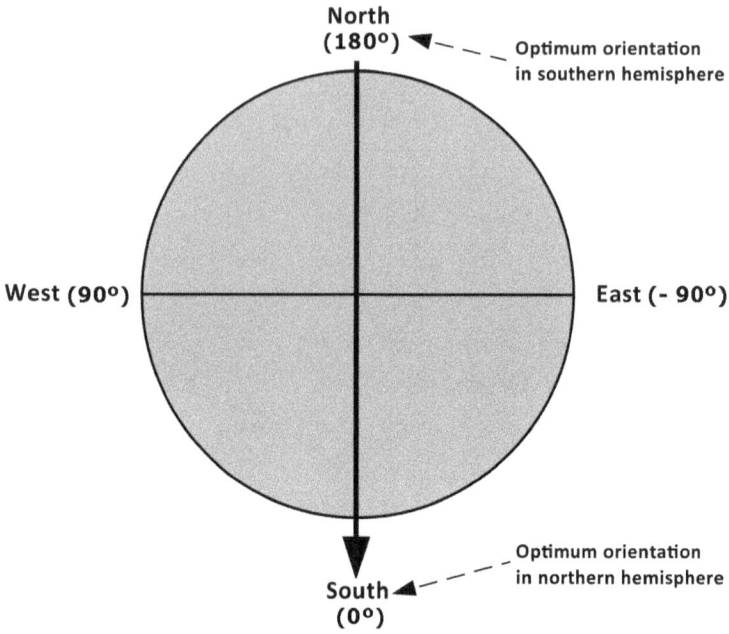

Figure 13: Module orientation in the northern hemisphere

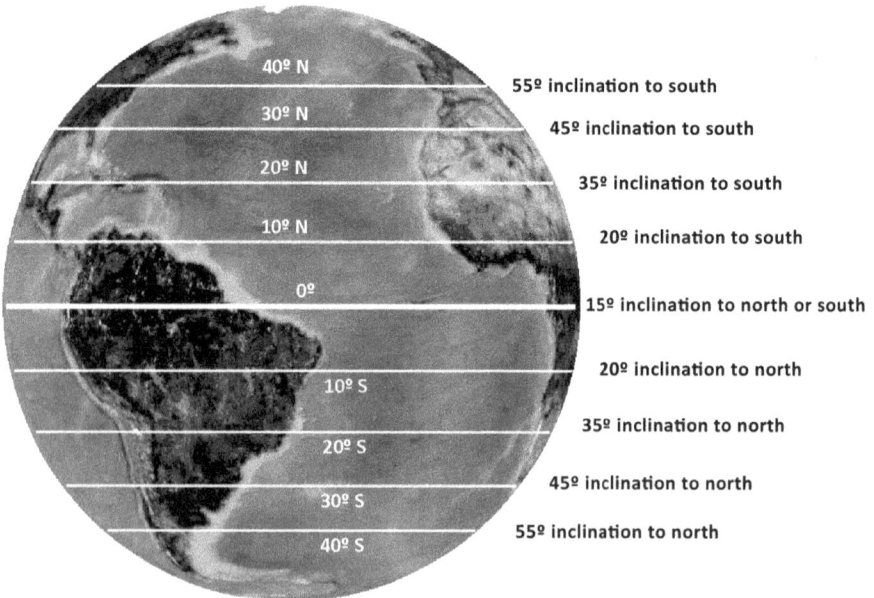

Figure 14: Recommended angle of inclination for modules according to latitude

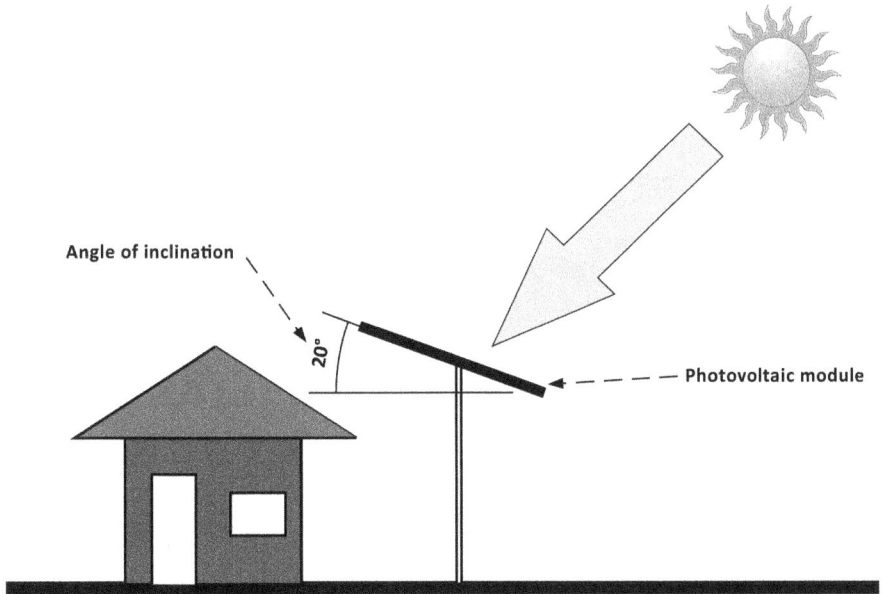

Angle of inclination

20°

Photovoltaic module

Figure 15: Recommended angle of inclination for latitude 10 º N/S

2.5 ORIENTATION AND ANGLE OF INCLINATION

For low power SPV systems, it's best to mount the modules at a fixed angle and orientation throughout the year. This simplifies the installation, reduces costs and makes maintenance easier. To maximize the energy collected, the modules must be orientated towards the equator. This means that for systems located north of the equator, the modules need to be orientated towards geographic South (0º). South of the equator, on the other hand, modules need to be orientated towards geographic North (180º). (Figure 13).

The optimum angle of inclination depends on a series of factors: the latitude of the site, the sizing method and the evolution of energy demand over the course of the year. Figure 14 shows the recommended angle of inclination for modules at different latitudes (assuming demand is constant throughout the year and the worst-month method is used to size the system; see Chapter 10). If you are at latitude 10 ºN, the angle of inclination should be approximately 20 º (Figure 15).

2.6 ENERGY EFFICIENCY

When we talk about energy efficiency, we are referring to the relationship between the energy input and the energy output of a system. When a system has an 80% efficiency, it means 20% of the energy entering the system is "lost" due to inefficiencies. In practice, energy can't be lost, it's simply transformed into other kinds of energy, which may or not be useful to us.

For example, a coal-fired electric power station has an energy efficiency of approximately 30%. This means that 70% of the energy entering the system is converted to other forms of energy that are not electricity (mainly heat). The following worked example shows an example of how to calculate power output of a PV module based on its efficiency:

$$\text{Efficiency (\%)} = \left(\frac{\text{Energy In}}{\text{Energy Out}}\right) \times 100$$

Due to the high initial cost of an SPV system, it's important to make the most of the energy the system produces. It is almost always easier and cheaper to save energy than it is to generate it. Sizing and installing an SPV system to light a house with 100W incandescent light bulbs (very inefficient) is like driving a car with the handbrake on, wasting a lot of the energy that the motor generates. The same amount of light can be generated using energy saving light bulbs.

Worked Example 6: Power output of a PV module based in its efficiency

If I have a module measuring 1m x 1m ($1m^2$) with an efficiency of 15%, what is the output power in the midday sun, when irradiance is $1000W/m^2$?

Output power (W/m^2) = input power (W/m^2) x efficiency (%)

Output power (W/m^2) = 1000 W/m^2 x 0.15 = 150 W/m^2

The price difference between an incandescent light bulb and an energy saving light bulb is relatively small, which means it is more economical to buy efficient light bulbs than to buy more solar modules to generate the 400 W extra that incandescent light bulbs will consume.

Energy efficient DC appliances for an SPV system (such as refrigerators) are expensive. However, in the long term, it can be cheaper than buying more modules and batteries. In any case, before calculating how much energy you need for your system, try to reduce demand by using efficient lighting systems and appliances. If you want to generate electricity for a 2000 W electric cooker, a 2500 W iron, a 5000 W arc welder and a 3400 W sound system with disco lights, it may be that an SPV system is not the best solution...

Worked Example 7: Energy efficiency of lighting systems

How much can I improve the energy efficiency of a lighting system if I substitute 5 incandescent 100 W light bulbs for 5 low-energy 20 W light bulbs?

Demand: 5 incandescent 100 W light bulbs = **500 W**

5 low-energy 20 W light bulbs = **100 W**

$$\frac{500 \text{ W} - 100 \text{ W}}{500 \text{ W}} \times 100 = \textbf{80\%}$$

By using energy savings light bulbs I can increase the energy efficiency of the system by **80 %.**

3 PHOTOVOLTAIC MODULE

The photovoltaic module is the generator in an SPV system. It converts the sun's energy into electrical energy via the photo-electric effect. A module is made up of a set of interconnected photovoltaic cells which, when exposed to the sun, generate an electric current.

The global production of photovoltaic modules has shot up in recent years due to growing demand, causing a significant fall in prices. The manufacture of photovoltaic cells is a hi-tech industrial process which was previously limited to developed countries. Nowadays modules are manufactured in a number of different countries, making stand-alone photovoltaic energy more accessible worldwide.

Figure 16: Let's see if this thing works!

Figure 17: Polycrystalline module

Figure 18: Monocrystalline module

The majority of photovoltaic modules currently available on the market can be divided into two technology categories: crystalline silicon and thin-film. Currently, crystalline modules are the most common, and are better suited to stand-alone systems. Within this category there are two types: monocrystalline and polycrystalline modules.

The cells of a polycrystalline module vary in tone and colour (Figure 17). You can tell monocrystalline modules apart because all the cells have the same colour and tone (Figure 18). Monocrystalline modules tend to have a higher efficiency than polycrystalline, although prices are generally similar. Photovoltaic cells are combined to form a module, and several modules together form an array (Figure 19).

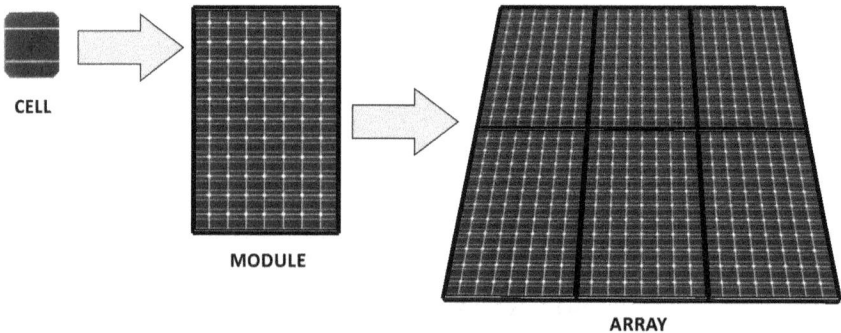

CELL

MODULE

ARRAY

Figure 19: Photovoltaic cell, module and array

3.1 ELECTRICAL CHARACTERISTICS

All modules are defined by their peak output in watts under what is called *Standard Test Conditions* (STC). STC is an industry standard which allows modules from different manufacturers to be compared under the same test conditions, being:

- ➢ **Solar irradiance:**   1000 W/m²
- ➢ **Cell temperature:**   25º C (77º F)
- ➢ **Spectral distribution:**   AM 1,5

The energy produced by a module will depend on 4 factors, which are:

- ➢ The available solar irradiance
- ➢ The angle of incidence
- ➢ The module cell temperature
- ➢ The voltage drawn by the load or battery

We'll have a look below at how these factors affect the energy produced by a module. To get a clearer idea of the module power output in real conditions, we refer to the *Standard Operating Conditions* (SOC), which are:

- ➢ **Solar irradiance:**   800 W/m²
- ➢ **Cell temperature:**   45º C (113º F)
- ➢ **Spectral distribution:**   AM 1,5

In this case the module power output is calculated based on a solar irradiance of 800W/m² and a more realistic cell temperature: 45º C (113º F), instead of 25º C (77º F). This is known as the *Normal Operating Cell Temperature*.

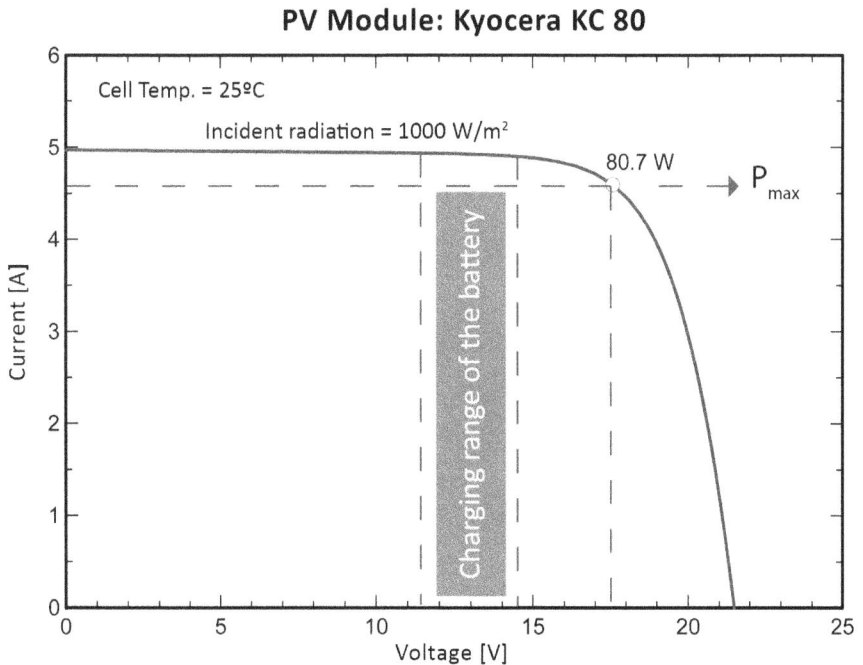

Figure 20: I-V curve for a kyocera KC80 module under standard test conditions (STC)

The power generated by a module under different conditions can be visualized more clearly with the current-voltage curve, or the *I-V curve*, which is available from the manufacturer. Figure 20 shows the I-V curve under Standard Test Conditions for a Kyocera KC80 photovoltaic module with a nominal power of 80W.

The vertical (Y) axis indicates the current (A). The horizontal axis shows the voltage (V). The I-V curve represents the output power of the module as a function of current and voltage. The *maximum power point* (P_{max}) is the point on the I-V curve just after it starts to fall steeply. In the diagram above, P_{max} is 80.8 W. If we follow the dotted lines from P_{max} we see that the voltage is around 17.6 V, and the current is around 4.6 A.

The grey rectangle shows the charging range of a 12 V battery (approximately 11 V – 14 V), if a non-MPPT charge controller is used (see Chapter 5). As we saw above, one of the factors which influences the energy produced by a module is the voltage required by the battery. In the diagram above, we can see that for a 12V SPV system, the module will be working at a voltage range lower than the voltage at the maximum power point (which is used by the manufacturer to define their module). Look at the I-V curve of the module you are thinking of using in your system – it'll help you to estimate more accurately the amount of energy the modules will produce. Remember that manufacturers sell their modules using figures for maximum power, which the modules will almost never generate in real life.

On the back of a module we find the manufacturer's label, showing the electrical characteristics of the module. Figure 21 shows the label of the module mentioned above.

In the first column of the table we can see the module's electrical characteristics under Standard Test Conditions (STC), and in the second column under Standard Operating Conditions (SOC). The figures are as follows:

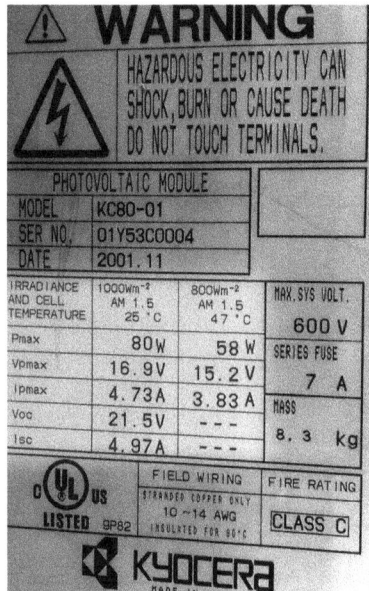

Figure 21: Identification label on a Kyocera KC80 module

➢ **P$_{max}$**: the maximum power or peak power of the module. Under STC this is 80 W, and under SOC it is 58 W (28% lower). If there are 8 modules in our generator, this will be a total of 176 W less than the nominal power under STC.

➢ **VP$_{max}$**: the maximum voltage or nominal voltage of the module under test and operating conditions. In other words, the voltage when the module is at its maximum power point, in this case 16.9 V under STC and 15.2 V under SOC

➢ **IP$_{max}$**: the maximum intensity (maximum current or nominal current) when the module is at its maximum power point. In this case: 4.73 A under STC and 3.83 A under SOC.

➢ **V$_{oc}$**: open circuit voltage. This is the voltage you would measure with a multimeter if you put the module in the sun, as it is the maximum voltage that the module can produce without any load.

➢ **I$_{sc}$**: the current intensity in short circuit. In other words, the maximum current that the module can generate when the voltage is zero (0 V).

Figure 22 shows the I-V curve of the module under Standard Operating Conditions (SOC). The I-V curve is different from the previous curve under STC, and the output power is lower: 58.5 W instead of 80.8 W (28% lower).

PV Module: Kyocera KC 80

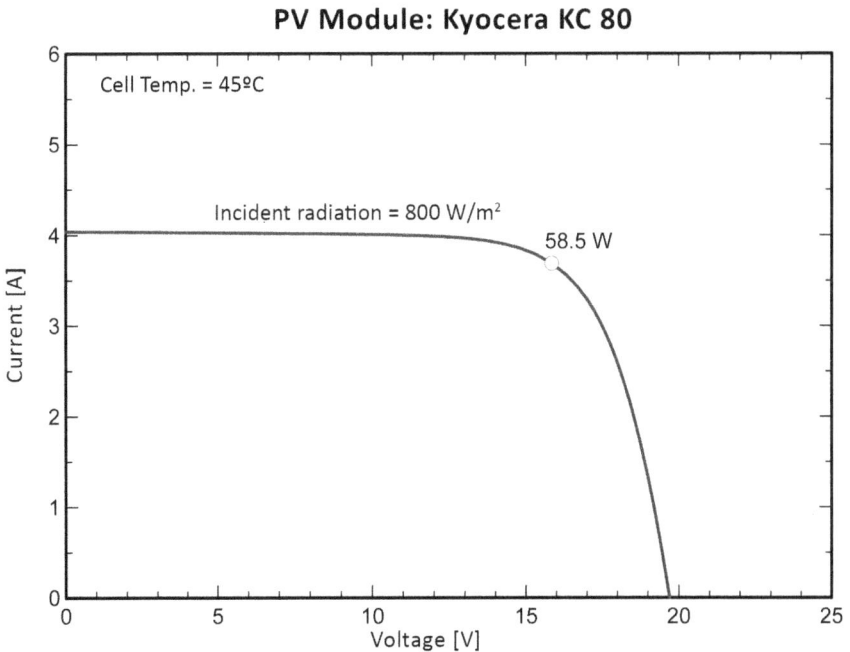

Figure 22: I-V curve of a Kyocera KC80 module under standard operating conditions (SOC)

3.2   EFFECT OF SOLAR IRRADIANCE ON MODULE POWER OUTPUT

To see how the power of the same module varies with different levels of solar irradiance, we look at Figure 23, which uses the same module as in the previous examples. The output power of the module is directly proportional to the level of solar irradiance. On a cloudy day with 400 W/m², P_{max} falls to 30.5 W.

PV Module: Kyocera KC 80

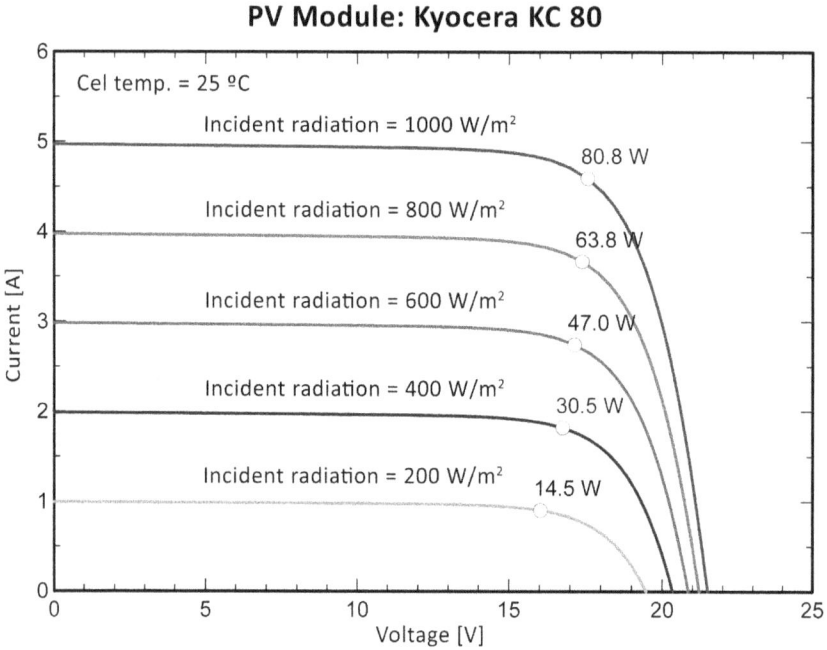

Figure 23: I-V curve of a Kyocera KC80 module (STC) with different irradiance levels

3.3   EFFECT OF TEMPERATURE ON MODULE POWER OUTPUT

The output power of a module diminishes by approximately 0.5% with every degree centigrade increase in cell temperature above 25º C (77 ºF). The temperature of cells in a module is normally 20º C (36 ºF) higher than the air temperature. If air temperature is 25º C, the cell temperature will be approximately 45º C (113 ºF). Figure 24 uses the I-V curve at different temperatures to show the effect of cell temperature on output.

If cell temperature is at 40 ºC (104 ºF) under standard conditions (SOC, 800W/m²), the power (P_{mpp} or P_{max}) falls to 58.5 W. If our SPV system is going to be installed somewhere where the air temperature is usually above 20 ºC, when it comes to sizing the system it's important to take into account the power loss due to temperature.

PV Module: Kyocera KC 80

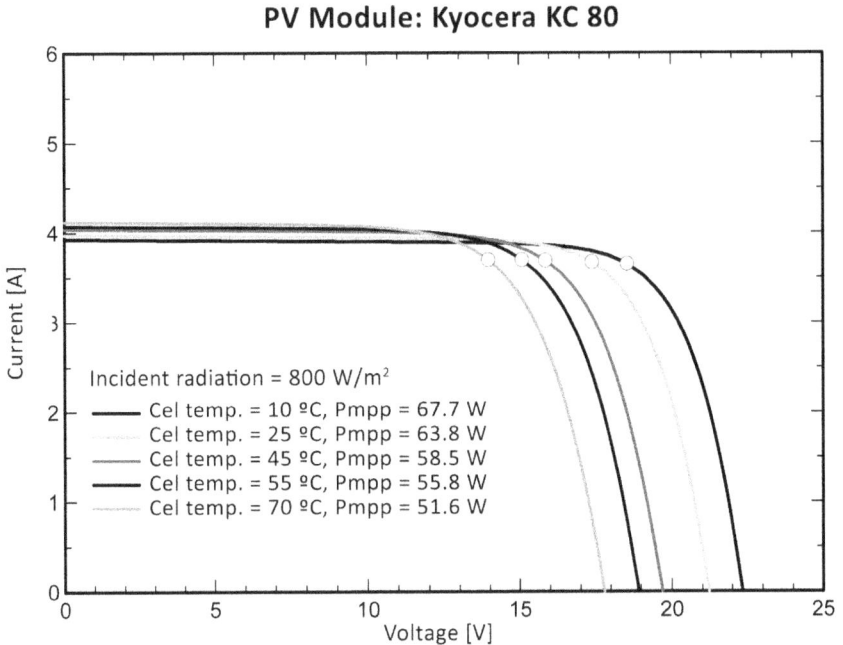

Figure 24: I-V curve of a Kyocera KC80 (STC) module at different temperatures

3.4   CHOOSING MODULES

The following list includes some of the important points to take into account when it comes to choosing and buying modules for an SPV system:

➤ **Module characteristics and application:** There is a wide variety of modules on the market: make sure that the module you are specifying is suitable for stand-alone systems and that its output voltage is right for the system voltage. Don't just trust what a manufacturer or supplier tells you. Check the module's I-V curve and specifications. If you are in doubt, consult an expert.

➤ **Warranty:** Make sure that the module comes with a guarantee. For crystalline modules, the warranty is normally between 5 and 10 years, and energy production at 80% of the nominal power is normally guaranteed for up to 25 years.

➤ **Cost/watt:** find the cost of different modules and work out the price per watt peak. An example is given in Table 4.

➤ **Buying second hand modules:** If you are thinking about buying modules second hand, test them with a multimeter in direct sunlight before you buy to make sure that the measurement you get corresponds with the data on the module label.

➢ **Module certification:** Check that the module comes with a label which includes the relevant data and an international certification, such as the IEC 61215 for crystalline modules.

Module	Price (US$)	÷	Pmax (W)	=	Price (US$/Wp)
Kyocera KD135GX-LPU	350	÷	135	=	1.81
Astronenergy 235	**360**	**÷**	**235**	**=**	**1.07**
Sharp 240	475	÷	240	=	1.39
AUO 250	755	÷	250	=	2.11

Table 4: Calculations of cost/watt peak for photovoltaic modules

Worked Example 8: Energy production of a photovoltaic module

I have a 12V SPV system with one Kyocera KC80 module and the solar irradiation at my location is 4 Solar Peak Hours per day. On average how much energy is the system going to produce each day in watt-hours?

Energy (Wh) = Current (A) x PSH. If I charge the batteries at 12.8 V, the I-V curve under Standard Operating Conditions (SOC) at 12.6 V is about 4 A.

4 A x 4 PSH = 16 Ah

16 Ah x 12 V = 192 Wh

In the example above, the Astroenergy 235 Wp has the lowest price per watt (1.07 US$/Wp), and costs approximately half that of the AUO 250 (2.11 US$/Wp). Check the module data sheet to work out the reason for a difference in price per Wp (it will primarily be to do with module efficiency). Make sure that it comes with certification and adequate warranty conditions. Remember that false economies can be dangerous!

4 BATTERIES

Batteries store the electrical energy generated by the modules during the day to be used by appliances during the night. They are usually the most sensitive part of an SPV system and require the most care and attention.

A battery bank will usually last for around 2 - 5 years, after which it needs to be replaced. This means the batteries represent the highest cost that end-users or operators will have to meet over the life cycle of the system. At the moment, lead-acid batteries are the most suitable for SPV systems: a technology with certain limitations but currently the best solution for stand-alone systems.

4.1 HOW A BATTERY WORKS

A lead-acid battery consists of one or more electro-chemical cells that convert electrical energy into chemical energy, a reversible process which means the battery can be charged and discharged in cycles. During this conversion process, approximately 25% of the input energy used to charge the battery is "lost" (converted to heat, mainly). There is a great variety of lead-acid batteries available (see Section 4.3). Figure 25 shows an example of sealed electrolyte gel batteries; Figure 26, shows flooded electrolyte liquid batteries.

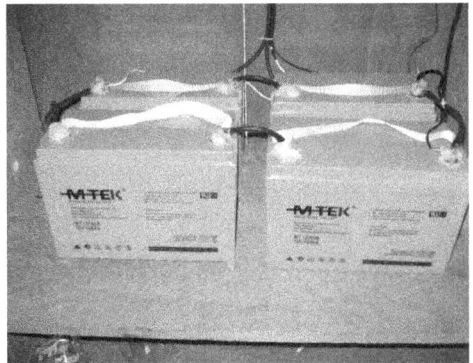

Figure 25: Sealed lead-acid batteries

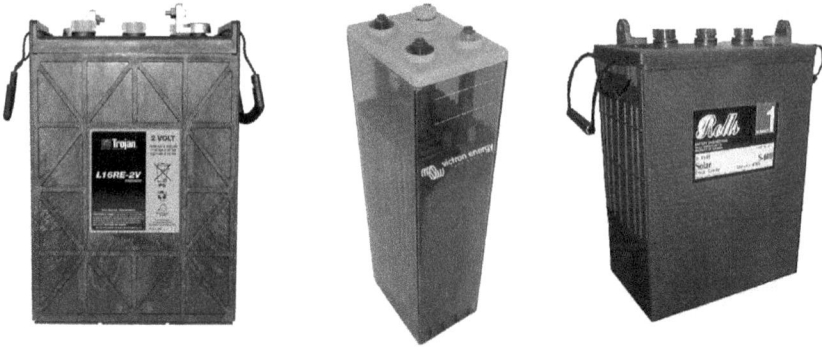

Figure 26: Flooded lead-acid batteries [source: Trojan, Victron Energy, Surrette]

The cell of a flooded lead-acid battery consists of lead plates submerged in an electrolyte of distilled water and sulphuric acid. The cells and electrolyte are enclosed in a plastic case with 2 terminals: positive and negative (Figure 27). Each cell in a lead-acid battery has a voltage of approximately 2.1 V when fully charged. Batteries are manufactured with a single cell (2V) or with several cells connected in series (6V, 12V, 24V). A fully charged 12V battery has a voltage of 12.6V when it is disconnected from any load. The voltage increases throughout the charge phase and decreases in the discharge phase. In order to charge a battery, the voltage charge must be higher than the voltage of the battery itself.

GASSING

Each cell in a flooded lead-acid battery has a ventilation cap, which lets hydrogen and oxygen escape during the charging process. This phenomenon is called *gassing*.

Gassing produces hydrogen, which is highly explosive. This means that the area where flooded lead-acid batteries are stored must be well ventilated and free from sparks. Put your pipe out or you may lose your beard!

SULPHATION AND STRATIFICATION

Sulphation is a phenomenon which affects a battery when it has been discharged repeatedly without being recharged, or when it has been left discharged for an extended period of time. When this happens, crystals accumulate inside the battery and its storage capacity is diminished. The main characteristic of sulphation is the appearance of greenish-blue coloured crystals on the positive terminal.

If caught in time, sulphation and stratification can be resolved by applying an equalisation charge. Equalisation consists of a controlled over-charge which causes intense gassing: the

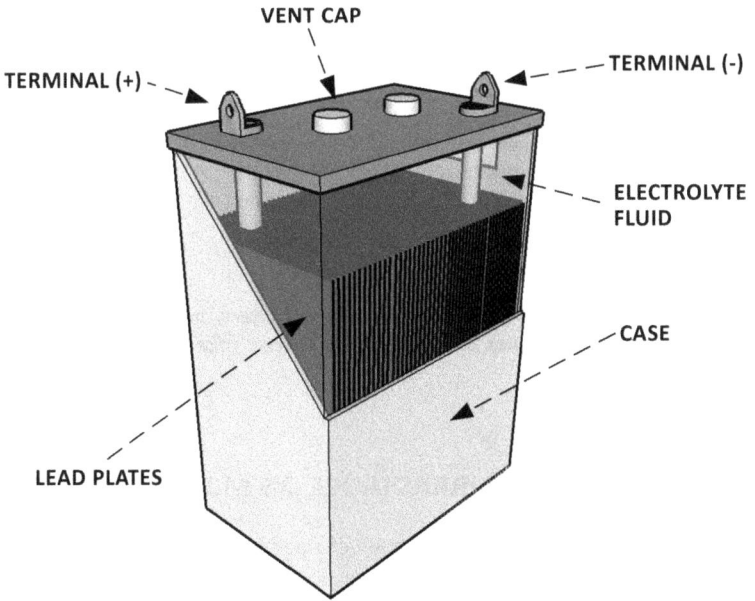

Figure 27: Components of a flooded lead-acid battery

Figure 28: A solar distiller

bubbles of hydrogen and oxygen stir up the electrolyte inside the battery and mix it up again, reducing sulphation and stratification effects.

REFILLING ELECTROLYTE

Because of gassing, flooded lead-acid batteries need to be refilled periodically with distilled water. Never use drinking water to refill electrolyte! You can buy distilled water from auto spare parts outlets, or make it on site with a solar distiller (Figure 28).

Do not burn or incinerate a used battery, or bury it near water sources, people's homes or crop plants. Always take used batteries to a recycling centre!

4.2 EFFECT OF SOLAR IRRADIANCE ON MODULE POWER OUTPUT

When a battery is discharged and then charged again, it is called a charge cycle. For SPV batteries, one charge cycle is generally the length of one day (24 hours). The amount of energy you get from the battery is called the *depth of discharge*, expressed as a percentage. If we discharge half of the battery's capacity, the depth of discharge will be 50% (Figure 29).

Figure 30 shows an example of the state of charge of a battery over the course of 7 days and nights, where the depth of discharge does not exceed 50%. Days 2 and 3 are cloudy, and the system does not succeed in replacing the energy used at night, such that the state of charge gradually falls to 50%. On day 5 the sun comes out again and the state of charge reaches 100 % again on day 7.

Depth of discharge and State of Charge

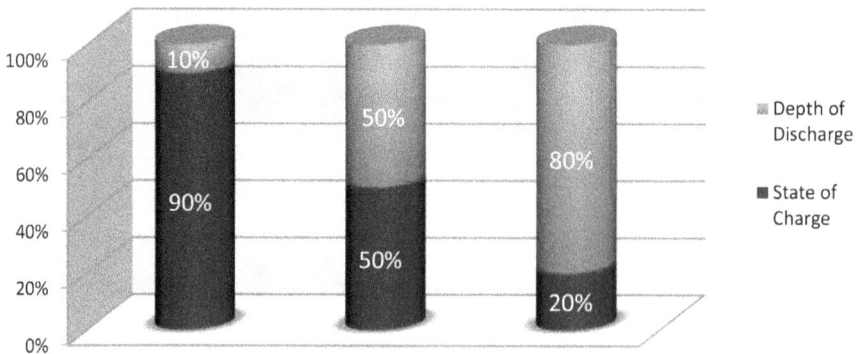

Figure 29: Depth of discharge and state of charge

Charge and discharge cycle

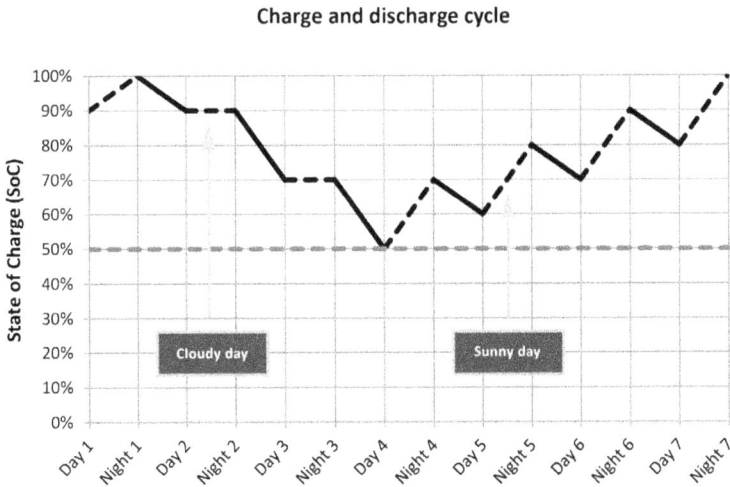

Figure 30: Charge-discharge cycle of a battery over a period of 7 days

CYCLE LIFE

The cycle life of a battery refers to the number of charge cycles it can undergo before its capacity is reduced to 80% of its nominal capacity, after which it needs to be replaced. The cycle life of a battery depends, among other things, on the depth of discharge of each cycle. Figure 31 shows the cycle life curve for a flooded lead-acid battery.

At a DoD of 50%, the battery will have a cycle life of approximately 1500 cycles. This is about 4 years (1500 ÷ 365 = 4.1). With a 10% depth of discharge, it would be approximately 6000 cycles (16 years), or 1000 cycles (2.7 years) at 80%. The depth of discharge directly affects the battery's cycle life. By way of comparison, a modified starter battery may have a cycle life of approximately 350 cycles (1 year) at a DoD of 50%.

As well as the depth of discharge, the charging current is an important factor that affects the cycle life of batteries. Low currents are better for charging a lead-acid battery (between 3 and 5% of the battery's charge capacity), and should never exceed a tenth of its capacity. For example, the charge current for a 100 Ah battery should never be more than 10 A.

There is a wide range of lead-acid batteries on the market, some more suitable for SPV systems than others. The different types of batteries are detailed in Section 4.3. Bear in mind that many manufacturers sell batteries with vague definitions like "solar batteries". Have a close look at the manufacturer's technical data sheets before making a decision.

Watch out for false economies when it comes to choosing batteries! If you go for the cheapest battery, it'll end up being more expensive in the long run.

Battery cycle life

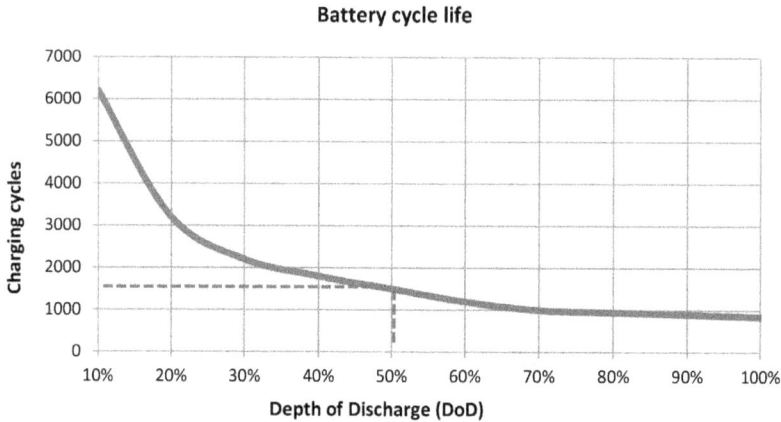

Figure 31: Cycle life of a flooded lead-acid battery

Worked Example 9: Cost analysis of batteries

I am deciding between two different batteries which I am thinking of using at a DoD of 50%. The first is a hybrid battery which costs US$ 200 and has a useful life of 1500 cycles (4 years) at a 50% DoD. The second is a modified starter battery and costs US$ 80. It has a useful life of 350 cycles (1 year) at the same depth of discharge.

After four years, which option works out to be the cheapest?

	Year 1	Year 2	Year 3	Year 4	Final Cost
Car battery	US$ 80	U$ 80	U$ 80	U$ 80	U$ 320
Hybrid battery	US$ 200	US$ 0	US$ 0	US$ 0	US$ 200

Every year, I will have to spend US$ 80 to replace the car battery. On the other hand, the hybrid battery will last for 4 years. At the end of the fourth year, I will have spent 320 US$ on car batteries and only US$ 200 on hybrid batteries. The hybrid battery represents a saving of US$ 120 after 4 years. (This is a simplified calculation that does not take into account that the value of money changes with time. See Chapter 11).

SELF-DISCHARGE

A battery left to stand without being charged gradually loses its charge, via what is known as the self-discharge effect (from 1 to 10% per week depending on the type of battery). The speed of self-discharge depends on the battery's composition, age,   condition, temperature, and how clean the surface is between the terminals. A dirty battery self-discharges more quickly, because current can flow through the dust and acid fog which accumulate on the surface between the terminals.

4.3 MEASURING THE STATE OF CHARGE OF A BATTERY

If you buy an aeroplane and the fuel gauge doesn't work, you could end up with no fuel mid-flight and you may have a rough landing. With batteries you need to measure the state of charge regularly to assess their condition. The voltage of a battery varies according to its state of charge. When a 12 V battery is fully charged, its voltage will be around 12.6 V (state of charge is 100%). With the state of charge at 50%, its voltage will be approximately 12.1 V. When the voltage gets to 11.5 V or less, the battery is dead. The electrolyte density in a battery also varies according to its state of charge: the more highly charged, the greater the density of the electrolyte. The density is measured in what is called the specific gravity of the electrolyte, in grams per litre, at a temperature of 25 ºC (77 ºF).

There are charge controllers and battery monitors that can indicate the battery state of charge in volts, some of which have an electronic function to compensate for the variations in battery voltage during charging or discharging periods. Aside from this, there are two ways to take an accurate reading of battery state of charge: using a hygrometer, which measures the specific gravity of the electrolyte in one cell of the battery; or using a multimeter, which measures the voltage of the whole battery. Table 5 shows the relationship between state of charge, voltage and specific gravity in a 12 V battery. For a 24 V battery bank, multiply by 2. For a 48 V bank, multiply by 4.

State of charge	Voltage	Specific gravity (g/l @ 25 ºC [77 ºF])
100%	12.73	1.277
90%	12.62	1.258
80%	12.50	1.238
70%	12.37	1.217
60%	12.24	1.195
50%	12.10	1.172
40%	11.96	1.148
30%	11.81	1.124
20%	11.66	1.098
10%	11.51	1.073

Table 5: State of charge of a 12V battery

HYGROMETER

The hygrometer is a device that measures the density of sulphuric acid electrolyte in each cell of a flooded battery. If you have a high quality hygrometer and a battery that has not become stratified, this is the most accurate way to ascertain the state of charge of a battery cell. Furthermore, it allows you to detect dead cells or cells which are about to die (something that you can´t detect using a multimeter).

A hygrometer consists of a glass tube with a heavy bulb of lead or mercury inside. It has a rubber bulb on one end, with a long rubber nozzle on the other end, with an opening to suck in the electrolyte and take the measurement. The heavy bulb is connected to a glass stem with a scale on it which allows the specific density of the electrolyte to be read directly. Some hygrometers only indicate a high, medium or low state of charge. Try to get one that has numbers on the scale so that you can take a precise reading (Figure 32).

To use a hygrometer, follow the steps below:

1.   Take the safety precautions mentioned above.

2.   Remove the ventilation cap of the battery cell from which you are going to take the measurement.

3.   Insert the long tube of the hygrometer into the battery cell until it is below the level of the electrolyte, and suck the electrolyte in, until the bulb inside is floating freely.

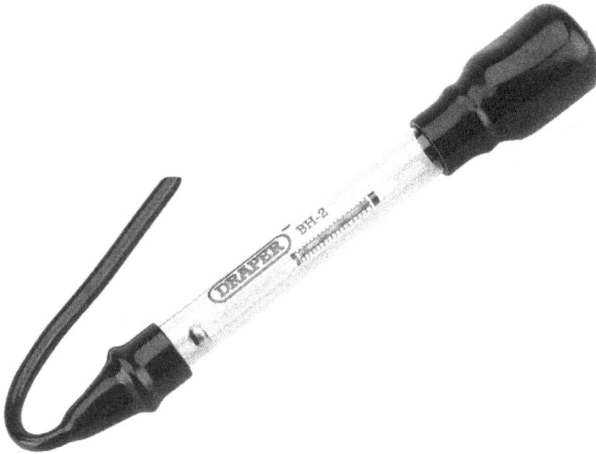

Figure 32: A hygrometer for flooded lead-acid batteries

4.   Look at the scale to see where the level of the electrolyte comes to and read off the number.

5.   Check Table 5 to see what the state of charge is for that cell.

Be careful when using a hygrometer! Wear protective glasses, gloves and old clothes. Make sure you have a source of fresh clean water nearby in case of acid spills. Clean the hygrometer carefully before putting it into the battery!

MULTIMETER

The multimeter is a key tool for any work related to SPV systems. It is a device that allows you to measure the voltage of a battery and work out its state of charge. Figure 33 shows a low-cost meter.

To take a reading with a multimeter, follow the steps below:

1. Make sure that the battery has been left to stand for 20 minutes without charging or discharging. Take the measurement first thing in the morning or disconnect the cables between the controller and the battery and/or between the battery and inverter.

2. Twist the dial on the multimeter until you have selected the position for measuring continuous current (DC or the symbol $\overline{}$), at an appropriate range (20 volts is the maximum for measuring a 12 V battery).

3. Connect the multimeter terminals (positive and negative) to the corresponding terminals on the battery.

4. Take the reading and check Table 5 to work out the state of charge of the battery.

Measuring the state of charge of a battery with a hygrometer has its risks and inconveniences. For larger SPV systems, taking readings with a hygrometer should be done by a qualified technician.

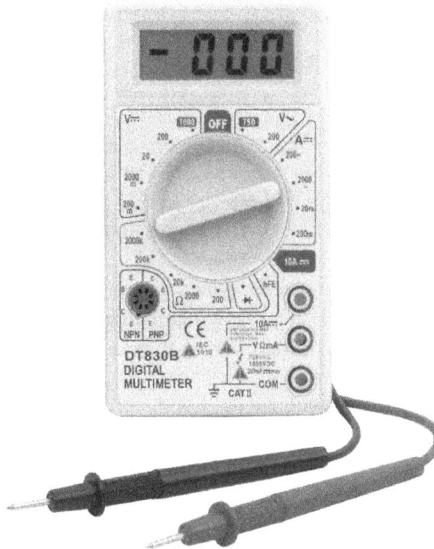

Figure 33: A low cost multimeter

Figure 34: Preparing to take readings with a multimeter

4.4 Charge capacity

The charge capacity of a battery for an SPV system is expressed in *amp-hours* (Ah). The capacity of a battery in Ah indicates the total amount of energy it can deliver before it is completely discharged. In theory, a 100 Ah battery can give a charge of 1A for 100 hours, 2 A for 50 hours, or 4A for 25 hours (1 x 100 Ah = 100 Ah; 2 x 50 Ah = 100 Ah; 4 x 25 Ah = 100 Ah).

In practice, the capacity of a battery varies with the speed of discharge (among other things). In other words, the same 100 Ah battery might give 100 hours of charge at 1 A, but only give 20 hours of charge at 4 A. This means that at 20 hours, the battery will have a nominal capacity of 80 Ah. For this reason, the charge capacity of a battery is always defined based on the discharge time, indicated by the letter 'C'. If a manufacturer indicates that its battery has a charge capacity of 100 Ah at *C 100*, it means that the capacity has been measured with a discharge time of 100 hours. C 20 indicates that the capacity is measured with a discharge time of 20 hours.

Which 'C' value do I use for choosing the batteries for an SPV system? In the majority of systems, batteries operate between C 100 and C 20, depending on usage. It is better to use C

20 for sizing battery banks, as it is more realistic for a low power SPV system (where a cycle is generally 24 hours: one day and one night) and is always less than the capacity in C 100 (which is what most manufacturers display on the batteries they sell). Amp-hours are a measure of charge, not energy. To find out how much energy there is in a battery, we multiply the charge in amp-hours by the battery voltage.

Worked Example 10: Converting charge to energy

If a 12 V battery has a charge of 50 Ah, how much energy does it have in watt-hours (Wh)?

Energy (Wh) = Charge (Ah) x Voltage (V)

Energy = 50 Ah x 12 V = **600 Wh.**

Effect of temperature on battery capacity

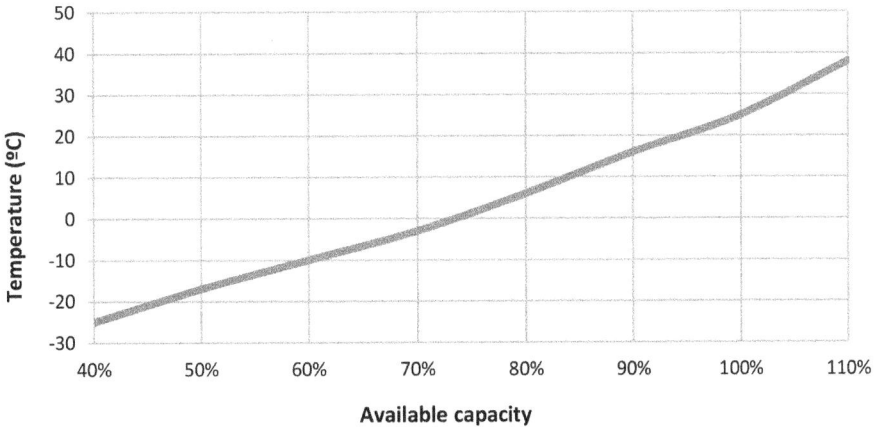

Figure 35: Capacity of a lead-acid battery as a function of temperature

Effect of temperature on battery cycle life

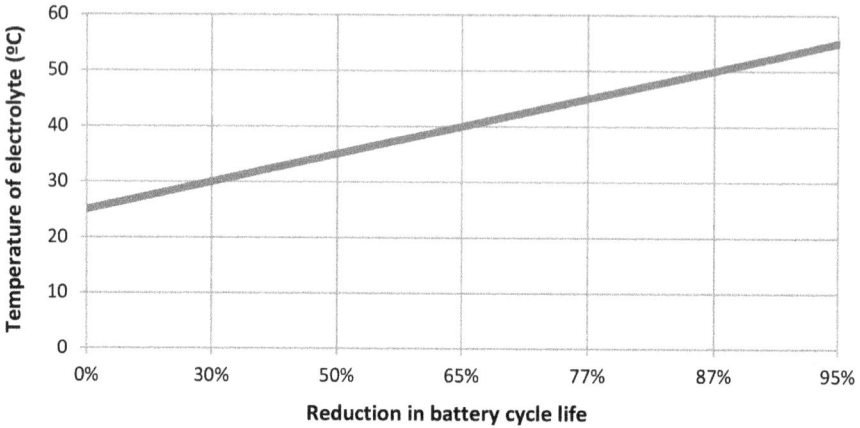

Figure 36: Reduction in the useful life of a lead-acid battery as a function of temperature

4.5 EFFECT OF TEMPERATURE ON BATTERY CAPACITY

The capacity of a battery also varies over time depending on its temperature – which is a factor you should bear in mind if your SPV system is going to be located where there are very low or very high temperatures. Figure 35 shows the variation in capacity of a lead-acid battery as a function of temperature.

The available capacity decreases as the temperature starts to fall below 25 ºC (77 ºF). At 0 ºC (32 ºF) the available capacity is reduced to approximately 75% of the nominal battery capacity. In places where low temperatures are common, housing the battery where it's protected from the elements where the ambient temperature is higher (i.e. inside a building) can limit these effects. For SPV systems in warm places, excessively high temperatures will be a more important factor. Figure 36 shows the reduction in useful life as a function of the electrolyte temperature.

The diagram shows that when the electrolyte temperature is higher than 25 ºC (77 ºF), the useful life of the battery starts to decrease. If the electrolyte temperature is at 30 ºC (86 ºF), the battery cycle life is reduced by 30 %.

4.6 TYPES OF BATTERIES FOR SPV SYSTEMS

There is a great variety of batteries on the market, which are categorised according to their usage regime or their composition. Table 6 summarises the main categories of lead-acid batteries and the commonly used names for each type:

Category	Common Names	Notes
Starting	Starting battery	*Generally not suitable for SPV systems*
	SLI (*Starting-Lighting-Ignition*)	
	Car / auto / truck battery	
Hybrid	Modified starter battery	**Suitable for SPV systems**
	Solar battery	
	Marine battery Leisure battery	
Deep cycle	Traction battery	**Suitable for SPV systems**
	Valve regulated battery (VRLA)	
	Gel battery (captive electrolyte)	
	AGM battery (captive electrolyte)	
	Tubular plate lead acid batteries, liquid or gel (OPzS o OPzV)	

Table 6: Lead-acid batteries

Be wary of what a manufacturer or retailer says about their batteries. Sometimes "maintenance-free" means that the battery will die just after the warranty runs out, and "deep cycle" or "solar battery" can mean many different things. Check the data sheet to work out what kind of battery you're really dealing with. Generally, price reflects quality, so if two batteries which are both called "deep cycle" are sold at very different prices, it probably means that the cheaper one is a hybrid battery and the more expensive is a true deep cycle battery. Sometimes the only way to be sure whether a battery is really deep cycle is to open it up. Obviously this has its disadvantages... If you are not sure, consult an expert in your area and read the manufacturer's data sheet.

There are many factors which affect the useful life of a battery, from the quality of its component parts, to how it's used. An important factor is the thickness of the lead plates inside the battery. The lead plates in a deep cycle battery are around seven times thicker than those in a car battery. Logically, this affects the weight. If you are comparing two batteries which are both sold as "deep cycle" (as in the previous example), look at the weight-volume ratio on the data sheet. This will give you an idea of which one has thicker lead plates.

In remote areas, there may be a very limited availability of deep cycle batteries, so you will have to decide whether importing batteries from abroad is justified in terms of costs and ease of replacement in the future. Use your judgment depending on the project you have in mind and make a decision based on the information given here. Below you will find summary characteristics of the most common batteries for SPV systems. The advantages and disadvantages of each battery are briefly explained.

STARTER BATTERIES

Starter batteries are the batteries used in automobiles. Their name is derived from their principle function: to start the engine, supplying a large amount of energy for a very short time, in very changeable temperature conditions. They have very thin, spongy lead plates.

Advantage
Starter batteries are usually the cheapest option and the most readily available on the local market. To make them work in an SPV system, you need to use them at a 10-15% DoD to prolong their useful life, and use truck batteries, which have a greater capacity.

Disadvantages
Generally, they are not suitable for SPV systems, so using them is a false economy. Unless you make sure the DoD is very low, they won't last long (6 months or less) as they aren't built for repeated deep cycling.

HYBRID BATTERIES

Often called "solar" or "marine" batteries. The electrodes in a hybrid battery contain a lead-antimony alloy, which allows for a greater depth of discharge. They contain thicker lead plates and have a greater electrolyte capacity than starter batteries.

Advantages
Hybrid batteries are definitively better than starter batteries, and often have a good price-quality ratio for low-power SPV systems. They are usually more widely available than deep cycle batteries. Their cycle life tends to be between 600 and 2000 cycles (2-5 years) if the depth of discharge is less than 25%.

Disadvantages
Their cycle life is reduced at deeper discharge cycles.

TRACTION BATTERIES

These are batteries that are generally used in a forklift truck. They contain thick lead plates, which are denser than those of starter batteries and have a high lead-antimony content.

Advantages
They tolerate deeper discharge cycles than starter batteries and hybrids. The antimony in the plates increases their useful life and resistance.

Disadvantages
Due to their high lead-antimony content they lose more electrolyte from gassing: as a consequence they require frequent refills of distilled water.

Tubular plate lead acid batteries with liquid or gel (OPzS and OPzV)

These have a long life at deep discharge cycles, and best suited to larger SPV systems. They come in voltages from 2-6V. The positive terminals are manufactured as tubular cells instead of flat plates, and protected by a tubular sleeve.

Advantages
These batteries are capable of withstanding very deep discharge cycles with a long cycle life. They usually have transparent casing which allows the electrolyte level to be seen easily.

Disadvantages
They are the most expensive batteries for stand-alone systems and are normally used in large installations, where the high price is justified by a long useful life at deep discharge cycles.

Valve regulated batteries (VRLA)

Valve regulated batteries (or VRLA, valve-regulated lead-acid) are sealed batteries. The name comes from the fact that, during the gasification process, the hydrogen recombines as electrolyte (water) inside the battery. If they over-charge, there is a safety valve that releases pressure and prevents extreme pressures inside the battery box. Some VRLA batteries are gel or AGM (see below). Those which are not gel or AGM batteries contain calcium in the lead plates in order to reduce the effect of gassing.

Advantages
In situations where maintenance needs to be minimized, they can be used in SPV systems if the discharge cycle is not very deep and the charge controller is adequate.

Disadvantages
To avoid electrolyte loss, they require a charge controller specific to this type of battery (see Chapter 5.) They are usually much more expensive than starter batteries and do not tolerate deep discharge cycles.

Gel batteries

Gel batteries contain sulphuric acid electrolyte converted to gel. They come factory-sealed.

Advantages
Because they are sealed, the electrolyte cannot be spilled, which makes transport simpler. They are zero maintenance and immune to stratification. Some can withstand deep discharges, with a life cycle of approximately 2 years at a 50% depth of discharge. They have a low rate of self-discharge.

Disadvantages
They are not suitable for warm climates, as the loss of electrolyte due to high temperatures can cause the battery to die prematurely. Many gel batteries do not tolerate deep discharges and are very sensitive to charge voltage, over-charging, and high charge currents. Because of these factors and their high price, they are not recommendable for SPV systems.

AGM Batteries

Absorbed electrolyte batteries (Absorbed glass matt, AGM) contain glass fibre mats in which the sulphuric acid electrolyte is absorbed.

Advantages
These batteries do not spill, are easy to transport and very robust. They can have a cycle life of around 4 years at a 40% depth of discharge, and have a very low self-discharge rate (1-3% per month). They do not emit hydrogen when charging and are immune to stratification. They differ from gel batteries in that they do not need a controller for specific charge voltages.

Disadvantages
They are expensive- two to three times more than flooded deep cycle batteries.

Stationary or Stand-by Batteries

These are batteries which are used as energy sources in an emergency.

Advantages
They have thick plates of pure lead, without antimony (they are usually very heavy).

Disadvantages
They are not designed for deep discharges, as they are generally kept fully charged, and are not very suitable for SPV systems. However, they can be used as long as the depth of discharge is never more than 30%.

Sometimes it is difficult to determine the typology of a battery without looking closely at the technical data sheet. On the battery in Figure 37, the label shows "Estacionaria (Stationary)...Deep Cycle". This kind of battery is sold in a rural area of Colombia where rural satellite telephones are connected to small SPV systems. This is not really a deep cycle battery, it is stationary or stand-by. However, given that it is available on the local market, it could be used in an SPV system as long as care is taken that the DoD is ≤ 30 %, and if equalization charges are applied and distilled water is refilled each month.

4.7 Choosing batteries

Which are the best batteries for my system? There is no magic answer: it depends on the system requirements, the budget available, maintenance, the manufacturing quality of the battery in question and its availability on the local market. Analyse these factors and look at the technical characteristics of each battery to work out its typology and make a detailed comparison. Ask for quotes from local suppliers to compare prices and performance.

In the absence of other options and if budget is very limited, truck starter batteries are a viable option as long as the depth of discharge is always ≤ 20% (in other words, as long as its state of charge is kept at 80% or more). In terms of cost and performance, hybrid batteries (traction, "marine" or "solar") are usually a good option for low power systems, as long as the

Figure 37: Stationary battery

depth of discharge is always 50% or less, and you take care to refill electrolyte with distilled water and apply equalization charges regularly.

In remote areas where transport is difficult and it is better to minimise maintenance tasks, sealed AGM batteries can work well, as long as the depth of discharge is always 40% or less. Their main drawback is their high cost. For more powerful systems, tubular plate liquid or gel batteries are the best option. The following list presents the important points to bear in mind when it comes to choosing and buying batteries for an SPV system:

➢ **Electrical characteristics:** try to find the technical data sheets or manufacturer's manuals for several different batteries to make a meaningful comparison. The battery is the most sensitive part of the system and the part that needs replacing frequently.

➢ **Do not buy used batteries:** it is very difficult to verify their true condition.

➢ **Availability on the local market:** it is recommendable to choose a battery that is available on the local market, to make it easier to replace in future. Start by finding out what batteries for SPV systems are available in your area and make a list of prices and characteristics.

➢ **Maintenance:** bear in mind that different batteries have different maintenance requirements. If it is unlikely that end-users or operators will check the electrolyte level and apply equalization charges, then consider spending more on sealed batteries.

> **Dimensions:** check the dimensions of the batteries to make sure that they will fit in the place where they are going to be installed, and that there is a suitable place which has sufficient thermal insulation and ventilation (for flooded batteries).

> **Capacity and useful life:** do not buy a battery without first sizing your system and having a clear idea of what is suitable for the needs and the usage it will get.

> **Cost per cycle:** a good way of comparing different batteries is to calculate the price per cycle. Table 7 shows an example, where the traction battery has the lowest cost, at 0.19 US$/cycle. The van starter battery costs more than double that: 0.56 US$/cycle!!

Battery	Price (US$)	÷	Useful life at 50 % DoD (cycles)	=	Price (US$/cycle)
Van Starter	100	÷	180	=	0.56
Hybrid	200	÷	900	=	0.22
Traction	**280**	÷	**1500**	=	**0.19**
AGM	400	÷	1900	=	0.21
Tubular	590	÷	3000	=	0.20

Table 7: Example of cost-per-cycle for different batteries

5 CHARGE CONTROLLER

A charge controller is located between the modules and batteries. It performs the following functions:

> **Protection against battery overcharging:** the controller regulates the energy coming from the module, depending on the battery's state of charge. When there is a lot of sun and the battery reaches full charge, the controller reduces the current flowing into the battery to protect it from over-charging.

> **Protection against battery over-discharging:** if energy demand is high and the modules are not able to replace the energy being consumed (at night, for example), the controller detects when the battery state of charge is too low and disconnects the load.

> **System monitoring and performance:** depending on the model, the controller can provide basic information about the state of the system, monitoring the voltage, current, and state of charge of the battery, via LED lights or an LCD screen.

As we saw in Chapter 4, batteries are highly sensitive to overcharging and over-discharging. Because of this, the controller plays a crucial part in the system, regulating the charge in the optimal way to protect the batteries. Fortunately, they are also one of the most reliable parts of the system and rarely give any problems. The rated voltage of the charge controller must correspond to that of the system, and its maximum rated current must be greater than the maximum short circuit current that the modules can generate (see Example Calculation 11).

Figure 38: A PWM charge controller [source: Steca]

5.1 CHARGE PHASES

Charge controllers usually charge batteries in 4 phases:

➢ **Initial charge:** when the battery voltage reaches a certain level, the controller allows all the available current from the module to flow in, until the battery reaches an 80% state of charge.

➢ **Absorption charge:** the controller maintains the final charge voltage but reduces the current coming from the modules until it reaches a 100% state of charge.

➢ **Float charge:** once the battery is completely charged, the controller only lets a very small current pass through to the batteries, to maintain the state of charge at 100%.

➢ **Equalisation charge:** a function in certain controllers, only to be used with flooded lead-acid batteries that have liquid electrolyte. The process consists of a controlled over-charge, to reduce acid stratification and sulphation inside the battery, as well as to level out the voltage in the different cells of the battery.

Some charge controllers allow you to select the type of battery in the system (liquid or gel electrolyte, for example) so that the voltage and duration of the charge phases are adjusted to suit the battery characteristics and optimize the charging process. This is important if you decide to use sealed gel batteries in your system: make absolutely sure that the controller has a charge setting for this type of battery, as they require lower charge voltages than other batteries.

5.2 TYPES OF CONTROLLER

Broadly speaking there are 4 types of controllers that can be used for SPV systems:

➢ **Series:** the simplest and cheapest controllers, these are generally not used in SPV systems nowadays. They are connected in series between the module and the battery. When the battery voltage reaches a certain point, they disconnect the charge. When the voltage falls again, the charge is reconnected.

➢ **Parallel (or shunt):** these are connected in parallel between the module and battery. They gradually reduce the charge current until the state of charge reaches 100%. They are cheap and simple and well suited to small SPV systems.

➢ **Pulse width modulation (PWM):** they emit pulses of charge to the battery. The "width" or intensity of the pulses depends on the battery's state of charge. The width of the pulses decreases progressively as the battery charges. It is a well proven technology and PWM controllers are usually cost effective and reliable.

➢ **Maximum power point (MPPT):** these have continuous current converters which enable them to work at the maximum power point of the module. As we saw in Section 3.1, the voltage at Pmax is usually much higher than the charging voltage of a 12 V battery. Because of this, an MPPT controller can increase the energy the modules receive by 10 – 35%. They are typically used in larger systems, where their cost (they can be up to twice the cost of a PWM controller) is justified by a higher yield from the generator.

5.3 ELECTRICAL CHARACTERISTICS

The key characteristics for charge controllers are as follows:

➢ **Nominal voltage:** the voltage at which the controller works, which corresponds to the system voltage. For low power SPV systems, the most common is 12 V. Controllers are available in 12 V, 24 V and 48 V. The controller must be able to work at the same voltage as the system.

➢ **Nominal intensity:** the maximum current supplied by the generator (and generally the maximum current that it can supply to the load). You need to calculate the maximum current from the modules and size the controller accordingly.

5.4 TEMPERATURE COMPENSATION

Some charge controllers have a temperature compensation function, adjusting the charge voltage according to the ambient air temperature (batteries require higher charge voltages at low temperatures). Others include cables that connect from the controller to a sensor mounted on the batteries, to optimise charging voltage according to the battery temperature.

> **Worked Example 11: Sizing a controller**
>
> If I have two 12 V modules connected in parallel that give a maximum current (IPmax) of 4 A, what will the nominal voltage and maximum current of the controller be?
>
> For connections in parallel, the current is added and the voltage stays the same, so the generator will give a maximum current of 8 A, at a voltage of 12 V (we will look at connections in series and in parallel later on). The controller needs to have the following characteristics:
>
> Current: **10 A minimum**; Voltage: **12 V**

5.5 CHOOSING A CONTROLLER

To select a controller for your system, follow these steps:

  ➢   **Read the technical data sheets and manuals** of different controllers for comparison and to help you make an informed decision.

  ➢   **Maximum current and voltage:** select the controller's maximum current and voltage according to the maximum current from the generator and the system voltage.

  ➢   **Monitoring the state of charge:** try to find a controller which shows the battery state of charge so that you can monitor the batteries.

  ➢   **Charge voltage:** if your batteries are sealed recombination batteries (VRLA or gel) make sure that the controller has a specific charge position for this type of battery.

  ➢   **MPPT:** if the budget allows, specify an MPPT controller, so you can take a voltage and current from the generator that corresponds to the maximum power point of the module. This will increase the amount of energy received from the modules will by 10-35 %.

6 INVERTER

An inverter is a device that converts direct current coming from the batteries to alternating current supplied to loads. The advantage of DC or continuous current is that it can be stored in batteries. However, most electrical equipment works with AC. In order to use these appliances with an SPV system we need an inverter. An inverter pushes up the cost of the system and reduces its efficiency. If your SPV system is just for lighting, you should design the system for DC only to reduce cost and losses.

A good solution for low power SPV systems is to have the lighting system connected to the charge controller in 12 V DC, with a small inverter for AC appliances.

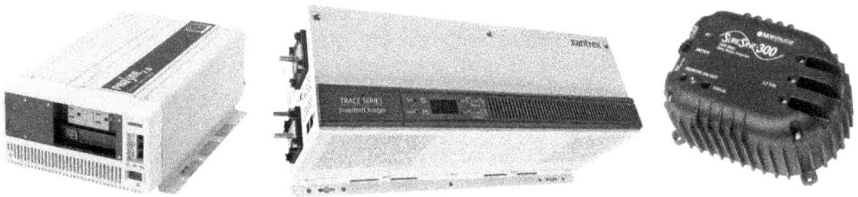

Figure 39: Inverters for SPVs [source: xantrex; morningstar]

6.1 TYPES OF INVERTERS AND HOW THEY WORK

An inverter contains a series of electronic circuits which convert low voltage DC to higher voltage AC. An inverter converts DC into AC and raises the voltage. Logically, the process of converting DC to AC entails losses, generally around 10 – 15 %. Most inverters have efficiencies of 85 – 90 % at full load. If used at lower loads (for example, a 1000 W inverter @ 500 W) efficiency can drop to 60%. These losses have to be taken into account when sizing the system.

Depending on which country you are in, AC voltage and frequency varies (the AC frequency is the number of times per second that the polarity changes direction, measured in Hertz). In most American countries AC voltage is from 110 – 127 V @ 60 Hertz. In Europe, Africa and Asia, voltage is generally 200 – 240 V @ 50 Hertz.

Inverters for SPV systems are defined by:

➢   Continuous power output, in watts (W), kilowatts (kW or kilovolt-amps (kVa). This is the maximum power that the inverter can supply, continuously.

➢   Peak power output: an inverter will be capable of supplying peak power for a very short time (generally anything from 10 seconds up to several minutes) for loads which have high starter currents, such as motors, compressors etc.

➢   Efficiency: generally from 60-95%; this varies considerably depending on the load. Inverters manufactured for use in automobiles have very low efficiencies and are not suitable for SPV systems (consult the efficiency curve from the manufacturer).

➢   DC input voltage: between 12 – 48 V for SPV systems.

➢   AC output voltage: 110 – 220 V, frequency 50 – 60 Hz.

➢   AC waveform, which indicates the "cleanness" of the current that is supplied to the appliances. These can be divided into square wave, modified wave and sinusoidal wave. Square wave inverters are usually the cheapest, but some appliances will not work with this type of wave. For appliances that are sensitive to wave quality, like computers or audio equipment, use sinusoidal wave inverters (they provide a cleaner AC wave, but are more expensive).

➢   Standby or no-load consumption: when all the AC appliances connected to the inverter are turned off, the inverter still consumes energy from the battery. Consult the manufacturer's data sheet to find out what this consumption is, and show the users or operators how to turn the inverter off when not in use. Higher quality inverters switch into rest mode when there is no load consumption, saving energy.

In low power SPV systems, it is important to minimise inverter usage, as it involves an additional cost and the losses in the AC part of the system are usually around 35% (compared to about 20% in DC). If a user starts to connect electric cookers, refrigerators, electric curling tongs and hairdryers just because there is an inverter, there will be problems! Efficient, low energy appliances need to be used with an inverter. Laptop computers, for example, consume half the amount of energy of desktop computers.

Watch out! Many low cost inverters are not suitable for SPV systems, as they are inefficient and do not have an appropriate waveform. Read the inverter datasheet or manual before making a decision.

Inverters range from just 50 W to thousands of watts. There are inverters for grid-connected photovoltaic systems and for stand-alone systems. Inverters for SPV systems can be divided into two general categories:

> Direct connection inverters: generally in the range of 150 – 2000 W, where AC appliances are connected directly to the inverter. They usually have between 1 and 3 output sockets.

> Cabled inverters: generally ≥ 1000 W, for connecting to a small AC network in a house or building.

There are also inverters with integrated chargers, which allow the batteries to be charged from an AC source (like an electric diesel or petrol generator). This kind of inverter is frequently used in back-up systems connected to the grid (Figure 40).

6.2 CHOOSING AN INVERTER

The following list includes some of the important points to take into account when it comes to choosing an inverter for an SPV system:

Figure 40: 1000 W sinusoidal wave inverter-charger

➢ **Efficient appliances:** try to use DC appliances wherever possible, leaving the fewest possible AC appliances that will be connected to the inverter. This will maximise system efficiency and minimise losses.

➢ **Inverter specifications:** get hold of the manufacturers' technical data sheets or manuals for different inverters, so that you can compare their quality-price ratios before you buy.

➢ **Continuous power:** work out how much continuous power you need based on the AC appliances (see Chapter 10 - Sizing). If the users turn on all the AC appliances at the same time and your inverter is under-sized, it will either turn off automatically or get damaged. By the same token, if you over-size your inverter, it will not be working efficiently, and you will be wasting money and energy.

➢ **Waveform:** find out what kind of appliances will be connected to the inverter so that you know which waveform is most suitable. Waveform-sensitive appliances will require a sinusoidal wave inverter.

➢ **Low cost inverters:** don't buy an inverter meant for automobiles or which has low efficiency, as it will consume a lot of energy and reduce the life of the batteries.

➢ **Standby consumption:** try to choose an inverter with a low standby consumption.

➢ **High/low voltage disconnect:** look for an inverter that has a function to automatically disconnect at low and high voltages, the former to protect the batteries from over-discharge and the latter to protect the inverter itself.

➢ **Used inverters:** be careful if buying a second hand inverter! Carry out tests with various different appliances connected; measure the output current, voltage and frequency, at full load.

7 LIGHTING AND APPLIANCES

Lighting and appliances are the last link in the chain of an SPV system, but that doesn't make them any less important! On the contrary: if they consume more energy than necessary, the system will be working inefficiently. Efficient appliances and lighting systems cost more than the less efficient alternatives, but overall costs for the system usually end up being lower, as you'll need fewer modules and batteries.

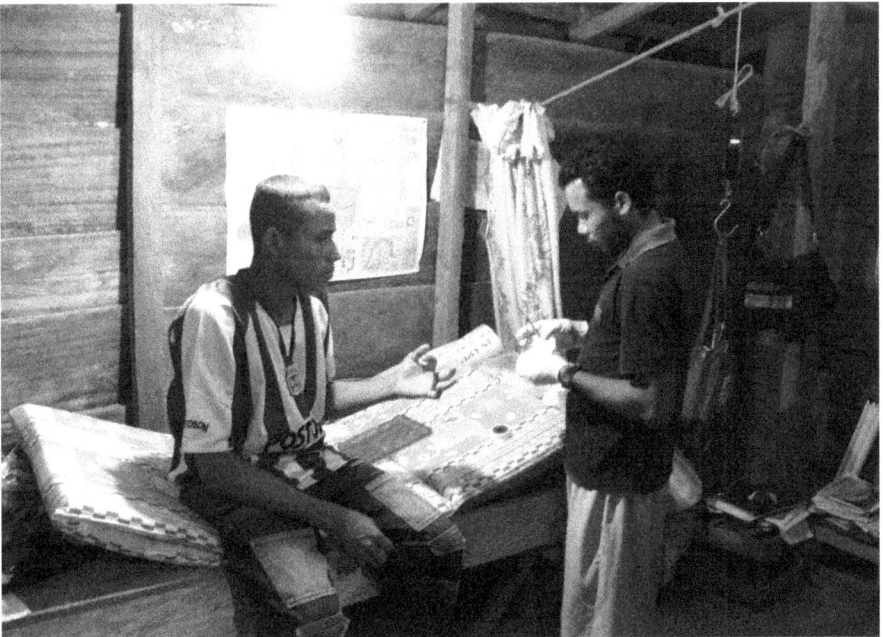

Figure 41: Lighting a rural health post with an SPV system

It's always better to reduce energy demand by using efficient appliances, than generating more energy to power inefficient equipment. This way the number of modules and batteries (which generally account for 50% of the initial cost of an SPV system) can be reduced, reducing overall system cost. There is little sense in spending a significant sum on an SPV system without taking into account the final amount of energy being consumed. Although most appliances and lighting systems on the market are AC, it is advisable, especially for low-power SPV systems, to use DC appliances and lights, to avoid inverter losses.

In the planning and design stages, identify which appliances are suitable for use, to guarantee the smooth functioning of the system and a long useful life for the batteries. It's important that the system users and operators understand which appliances should not be used, to avoid draining the batteries or damaging the system components (see Appendix 16.3). This chapter presents a summary of the lighting systems and appliances that are most suitable for low-power SPV systems.

7.1 LIGHTS

The electric light is an invention that has changed the world and improved the health conditions and wellbeing for millions of people. Providing light from a kerosene lamp, candles, or straight from an open fire, means emitting noxious gases which affect the health of the people involved and provide a poor quality of light.

The most common electric light bulbs for an SPV system can be divided into three categories: incandescent, fluorescent, and LED (light emitting diode). Each category is described in more detail below. The distinguishing features are their electrical power, their DC or AC voltage, the amount of light they emit, their colour temperature, and their useful life.

To decide which type of light is most suitable for an SPV system, understanding some basic concepts related to lighting and photometry comes in handy. Photometry is the measurement the brightness of the light perceived by the human eye, and is a useful tool for working out how best to light a given space, according to the activity of that space. For example, a school classroom requires a light with different qualities from that needed in the living room of a house. The most important parameters for lighting elements are summarized in Table 8, These will be useful when it comes to deciding which kind of light bulb is most efficient in terms of brightness, cost and energy, for illuminating a given space.

A kerosene lamp has a light output of approximately 100 lumens. A 60W incandescent light bulb produces approximately 850 lumens, with a luminous efficacy of 14 lm/W and a useful life of 1000 hours. A 3 W LED light bulb produces 240 lumens, with a luminous efficacy of 80 lm/W and a useful life of 50,000 hours. LEDs are much more efficient and have a much longer useful life!

Parameter	Description	Unit	Abbreviation
Luminous flux	The perceivable light output produced by a light source	*lumen*	lm
Illuminance	The amount of light output that falls on a unit of surface area, in lm/m^2	*lux*	LUX, o lx
Nominal power	The electrical power of a light	*Watt*	W
Luminous efficacy	The relation between the luminous flux of a light and its power	*lumens per watt*	lm/W
Colour temperature	The temperature of an ideal black body radiator that radiates light of a comparable hue to that of the light source	*kelvin*	K
Useful life	The average useful life in hours of a lighting system	*Hours*	h

Table 8: Parameters for lighting elements

The colour temperature of lights varies from a "warm" light (2700 K to 3000 K) to a "cold" light (4100 K to 6500 K). A candle has a colour temperature of approximately 1850 K. For the living room of a house, a "warm" light would be preferable (≤3000) K). On the other hand, for an office or workshop a "cold" light would be better (≥ 4000 K). The light we get from the sun at midday on a cloudless day has a colour temperature of approximately 20,000 K. On a cloudy day it falls to around 6500 K.

Table 9 shows various examples of different lights and their characteristics. Apart from the colour temperature, the most important characteristics when it comes to choosing a light for an SPV system are the luminous efficacy and useful life:

Category	Type	Nominal power (W)	Luminous flux (lm)	Luminous efficacy (lm/W)	Colour temperature (K)	Useful Life (hours)
Combustion	Candle	-	-	0.13	1850	20
Incandescent	Incandescent tungsten	100	1200	12	2700	1000
	Halogen tungsten	100	2500	25	3000	2000
Fluorescent	Compact fluorescent	20	1200	60	5000	8000
	Tube with electronic ballast	40	3200	80	5000	8000
LED	LED light bulb	3	240	80	*Varies*	50000

Table 9: Typical characteristics of different lights [source: McMullan]

The electrical and light characteristics of each type of light vary considerably depending on the manufacturer and the model. Incandescent, fluorescent and LED lights come in a variety of colour temperatures for different applications, in DC as well as AC.

It generally makes more sense to use DC lights with a low-power SPV system, as it improves the energy efficiency of the system (without the losses caused by an inverter) and simplifies remove installation and maintenance.

7.2   NATURAL LIGHTING

Before deciding which artificial lighting systems are most suitable for your SPV system, make the most of natural light inside the building so that electrical energy use for lighting can be reduced during the day. Skylights, light tubes or sun tubes are effective ways of getting natural light into a building (Figure 42 and 43).

The low-cost "solar light bulb" shown above, lets daylight into the building during the day, meaning artificial lighting can be avoided (developed by students at the Massachusetts Institute of Technology and the Myshelter Foundation). This system makes use of recycled PET bottles, filled with water and a small amount of bleach, sealed where it penetrates the roof for waterproofing.

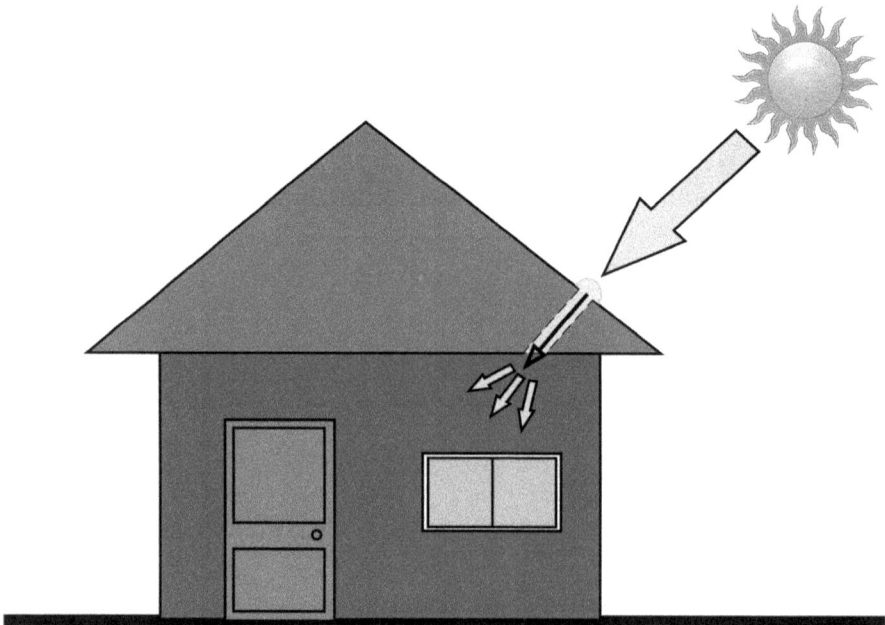

Figure 42: Light tube for making use of natural light inside a building

This kind of passive lighting can provide adequate light for a windowless space during the daytime, at zero operating cost. Another passive way to make a space brighter is to do with how reflective the surfaces are, for example the walls and ceiling. Dark surfaces absorb more light, so a room where the walls and ceiling are painted white will be brighter. It is better to repaint the surfaces of a space with white paint every two or three years rather than increasing the power of the lights.

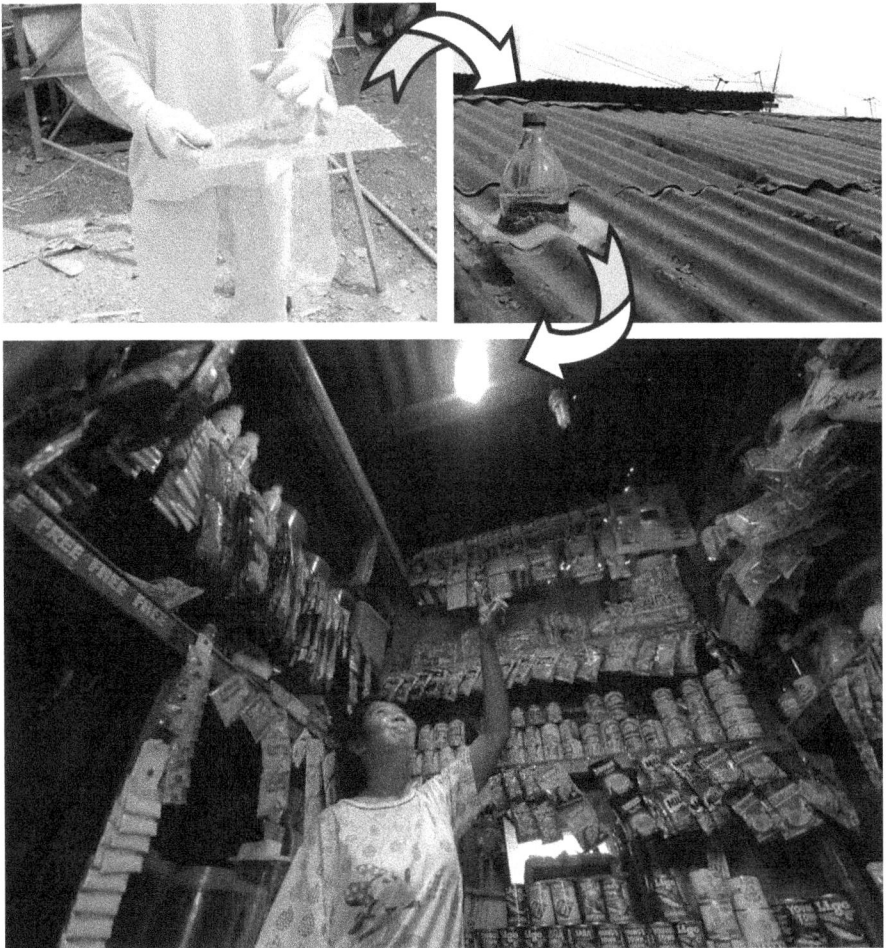

Figure 43: Solar light bulbs for low-income homes [source: Myshelter Foundation]

7.3 INCANDESCENT LIGHT BULBS

The incandescent tungsten light bulb was invented by Thomas Edison in 1879 and has become the most widely commercialised light bulb across the world. It consists of a metallic filament made of tungsten, enclosed inside a glass bulb which contains an inert gas. When the electric current passes through the filament, it glows and emits light. The inert gas stops the filament from burning, as it reaches temperatures of up to 3500 ºC (6332 ºF).

Figure 44: An incandescent tungsten light bulb

Nowadays, incandescent lighting is a technology nearing extinction, due to its high inefficiency: 90 % of the energy it consumes is turned into heat, and only 10% into light. It offers very low luminous efficacies (≈ 12 lm/W; see Table 9) and a very short useful life (1000 hours). In 2011 there were approximately 12,000 million electric light bulbs in use worldwide. If we only used efficient light bulbs we could save enormous amounts of energy. In certain countries, it is now illegal to sell this kind of light bulb because of its energy inefficiency.

Halogen light bulbs are a variation on the incandescent tungsten light bulbs. They incorporate a small amount of halogen gas in the bulb, which is made of quartz instead of glass, allowing the filament to operate at higher temperatures. Compared to tungsten light bulbs, their luminous efficacies are up to 50 % better (≈ 25 lm/W), they have greater lighting potential at a smaller size, and their useful life is approximately twice as long (≈ 2000 hours).

Due to their high inefficiency, it is not advisable to use incandescent light bulbs with an SPV system. If you decide to use them, try to use them only in rooms where they will be used for very short stretches of time (store rooms, for example). The only advantage of incandescent tungsten bulbs is that they have a low initial cost and are widely available. However, this is generally a false economy, as their useful life is much shorter than the more efficient alternatives and they consume far more energy.

Figure 45: Incandescent halogen light bulb

7.4 FLUORESCENT LIGHT BULBS

Fluorescent light bulbs consist of a tube, whose inside surface is covered with a fluorescent layer, filled with a vapour consisting of mercury and argon or krypton gas. As the electric current flows, the vapour emits ultraviolet radiation, which is converted to visible light as it comes into contact with the fluorescent layer on the tube. The composition of this layer can be adjusted to change the colour temperature by between 3000 and 6500 K. This means that fluorescent light bulbs can produce anything from a "warm" light to a very "cold" light, depending on how they are made.

Fluorescent light bulbs can be either compact or tube, and have efficacies of between 60 and 80 lm/W: at least four times greater than incandescent tungsten light bulbs, with a long useful life (8000 hours). They are well suited for applications where artificial lighting

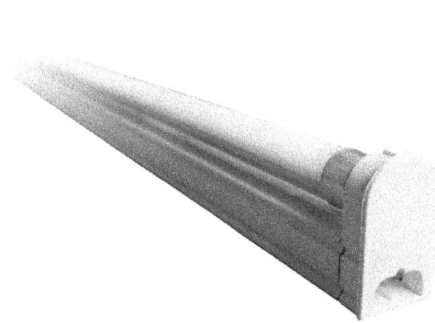

Figure 46: Fluorescent tube light   Figure 47: Compact fluorescent light [source: Steca]

is required for long periods of time, like classrooms, offices, commercial warehouses and kitchens. Given their high efficiency, long useful life and low cost, they are well suited to SPV systems. Both compact fluorescent and fluorescent tubes are available in DC.

This type of direct current light consists of a light bulb, a support structure and an electronic ballast. This last part does the job of inverting DC to AC, and transforming the current, as fluorescent light bulbs work with AC at 70 – 100 V. Figure 46 and Figure 47 show a fluorescent tube light and a compact fluorescent light.

Although compact bulbs are more expensive than tubes, they are generally better suited to SPV systems: they are easier to install (as they can be inserted directly into light sockets or standard sockets) and are often more efficient than tubes.

By way of comparison Table 10 shows the nominal power of compact fluorescent light bulbs and incandescent tungsten light bulbs. The light bulbs in each group emit approximately the same light output. The last column of the table shows the improvement in energy efficiency:

Type of light bulb	Nominal power (W)			
Incandescent tungsten	40 W	60 W	75 W	100 W
Compact fluorescent	9 W	11 W	15 W	20 W
Improvement in energy efficiency (%)	78 %	82 %	80 %	80 %

Table 10: Energy efficiency of incandescent and compact fluorescent lamps
[source: McMullan]

An energy saving of up to 82% can be achieved by using compact fluorescent light bulbs instead of incandescent. Their cost is higher than the incandescent bulbs but is justified by their long useful life and low energy consumption. The disadvantages of fluorescent light bulbs lie in their mercury content and the fact that it is not possible to regulate the light intensity.

Fluorescent light bulbs contain mercury, which is highly toxic. When they reach the end of their useful life, take them to a recycling centre or make sure that they are not disposed of near water sources.

7.5   LEDs

LEDs (light emitting diodes) employ semi-conductors to generate light without the need for gases or filaments, and have none of the mercury present in fluorescent light bulbs. They offer very high efficiencies (≥ 80 lm/W) and an extremely long useful life (≈ 50,000 hours). The development of LEDs in recent years means that the technology is very well suited to lighting systems for SPV systems. They are available in a wide range of colour temperatures, and the light intensity can be regulated. As a single diode does not normally generate sufficient light, there are usually several LEDs mounted together on the same fixture (Figure 48).

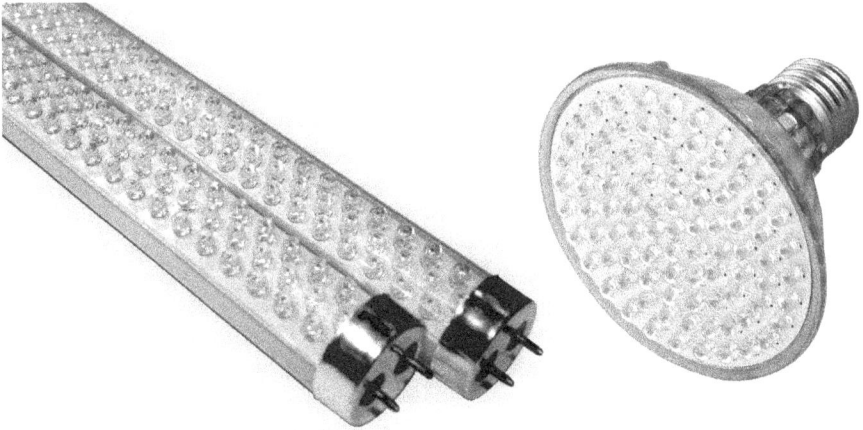

Figure 48: Examples of LED light bulbs

The fact that multiple LEDs are built into one tube means they can be installed in existing brackets for fluorescent tubes (image on the left of Figure 48). Currently their main drawback in their high cost.

7.6 CHOOSING LIGHTING SYSTEMS

The following list includes some of the important points which should be taken into account when it comes to choosing and buying lighting systems for an SPV system:

- ➢ **Application:** establish what application the light system is needed for (working, studying, reading, leisure etc.) and if direct or diffuse lighting is required. Choose a lighting system to meet the needs of the space you want to light.

- ➢ **Voltage and current:** DC light bulbs are generally more efficient than their AC counterparts. A DC lighting system will improve energy efficiency and simplify the system. If an inverter is present for AC loads, there will still be lighting available in case of inverter failure.

- ➢ **Luminous efficacy:** choose light bulbs with a high luminous efficacy to reduce energy use.

- ➢ **Useful life:** light bulbs with a long useful life will reduce operating costs for the system users or operators.

- ➢ **Availability on the local market:** take into account whether the light bulbs are locally available, so that users or operators can easily get hold of replacements.

> ➢ **Cost:** weigh up the above criteria together with the cost of different types of lights and the available budget for the system. Bear in mind that if you specify a light bulb with insufficient light output to reduce costs, it is possible that in the future the users may swap it for an incandescent light bulb with a higher light output, increasing energy use and reduce the efficiency of the system.

> ➢ **Light fixtures and brackets:** think about the brackets and fixtures, how they are mounted and how they direct light to where it´s needed. Light bulbs installed outdoors need protective covers to prevent moisture and insects from getting in.

7.7 APPLIANCES FOR SPV SYSTEMS

The appliances that can be used with an SPV system can be split into two broad categories: DC or AC. In general, DC appliances are more efficient than AC. For a low power SPV system, appliances which have electrical resistors (like electric cookers, irons, or heaters) or AC electric motors (fans, compressors, drills, saws etc.) should be avoided. It is better to use only laptop computers as opposed to desktops, as laptops use approximately 30% less energy. The typical power consumption of common appliances and their suitability for SPV systems can be found in the table in Appendix 16.3.

Manufacturers and suppliers of appliances for photovoltaic systems will often have appliances such as refrigerators and televisions available in DC, as well as DC adaptors for charging mobile phones directly from a 12 V source. The disadvantage is their high initial cost

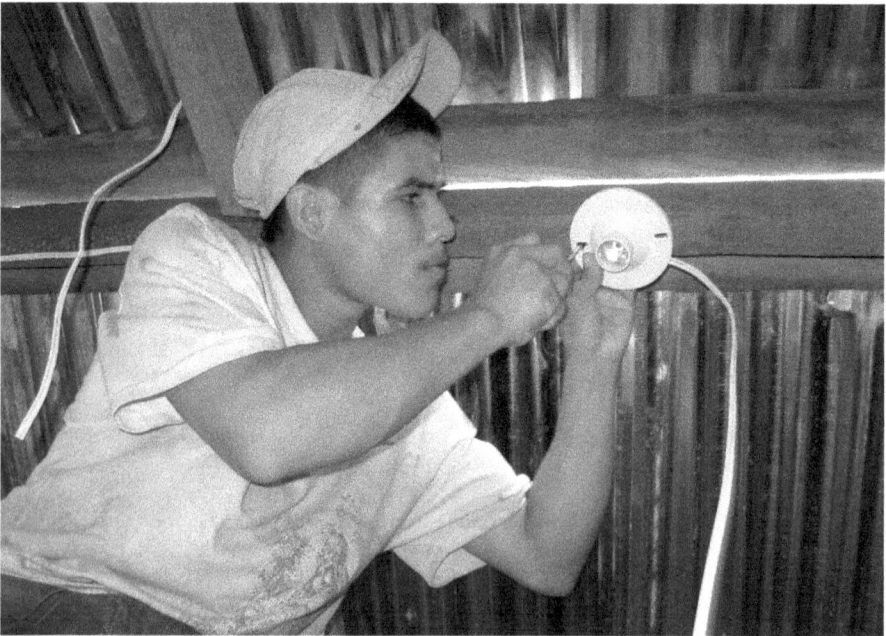

Figure 49: Connecting lights for an SPV system

and limited availability in remote areas. However, in the medium term, investing in efficient appliances can be more economical than buying more modules, batteries and an inverter for AC appliances.

In the planning phase of an SPV system, draw up an audit of the appliances that will be used now, and in the future, with their nominal voltage and power consumption. This will help to determine the system voltage and whether or not an inverter is required. Bear in mind that using an inverter in an SPV system makes the system more expensive, reduces energy efficiency and increases the number of complex appliances that can go wrong. In remote places, finding a specialised technical service for inverter repair can be difficult and costly. Include an inverter in the system only if the following conditions are met:

> AC appliances are efficient.

> There is already an AC electrical installation in the building.

> End users want to use AC appliances that cannot be replaced or adapted for use with direct current.

7.8 CHOOSING APPLIANCES

The following list includes some of the important points to take into account when it comes to choosing and buying appliances for an SPV system:

> **Efficient DC appliances:** DC appliances are in many cases more efficient than AC. If the budget allows, specify DC appliances, made specifically for photovoltaic systems.

> **Computers:** laptop computers use around 30% of the energy a desktop computer uses. Generally, computers require a sinusoidal wave inverter to work properly.

> **Refrigerators and freezers:** DC fridges and freezers can cost two or three times more, but they consume less than 25% of the energy of their AC equivalents. For rural clinics, it is best to buy a dedicated kit for solar refrigeration for medical applications. If AC refrigerators are used, they must be highly energy efficient and preferably the door should be on top, so that when the door is opened they lose less cold and consume less energy.

> **Mobile phones:** mobile phone manufacturers supply DC chargers. Low cost DC chargers made for automobiles are usually very inefficient.

> **Office equipment:** look at the data sheets of the equipment to find their power ratings and the waveform required, to work out what kind of inverter is needed.

> **Televisions:** DC televisions are very efficient but are usually hard to find. If AC televisions are used, avoid plasma screens and look for LCD or LED screens, which are more efficient.

8 WIRING AND TERMINALS

The wiring in an SPV system is equivalent to the arteries and veins of the human body: they transport energy between the different parts of the system. Sizing cables correctly is important to ensure that the system works efficiently and has a long working life. This chapter presents a summary of the methods to follow when specifying the cabling in a low power DC photovoltaic system (≤ 500 Wp). For any electrical installation you should follow the national codes and the electrical regulations applicable in your country.

8.1 CABLING

When electrons pass through a cable, part of the energy they carry is turned into heat, which represents losses in our SPV system. The two things which have an impact on these losses are:

> **the current,** measured in Amps (A).

> **the resistance of the cable,** measured in Ohms (Ω), a function of its cross-sectional area and material

In order to minimise wiring losses in an SPV system, we need to make sure that the cables have a suitable cross-sectional area: the greater the size of a cable, the lower its resistance. Using cables with a greater cross-sectional area implies higher costs, which is why you need to minimise the distances between components. This saves both money and energy.

Another way of reducing costs and losses in the cabling of an SPV is to increase the system voltage, which reduces the current running through each circuit. This is why larger SPV systems usually have voltages of 24 V or 48 V. Following the Power Equation, if we reduce current and increase voltage by the same factor, power will remain the same. Let's look at an example of three DC circuits with the same power (Figure 50).

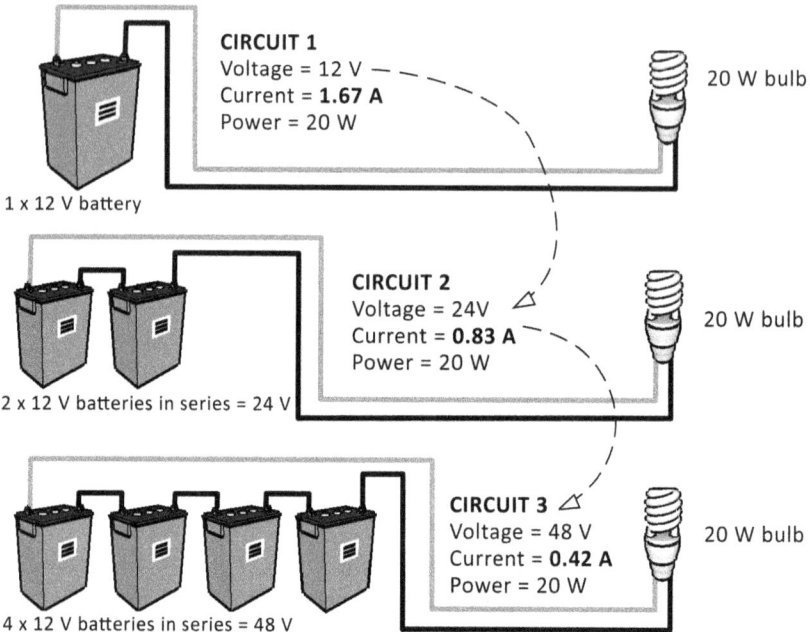

CIRCUIT 1
Voltage = 12 V
Current = **1.67 A**
Power = 20 W

20 W bulb

1 x 12 V battery

CIRCUIT 2
Voltage = 24V
Current = **0.83 A**
Power = 20 W

20 W bulb

2 x 12 V batteries in series = 24 V

CIRCUIT 3
Voltage = 48 V
Current = **0.42 A**
Power = 20 W

20 W bulb

4 x 12 V batteries in series = 48 V

Figure 50: The effect of voltage on current in 3 DC circuits

Circuit 1: the circuit voltage with a single battery is 12 V, and the power of the light bulb is 20W. According to the Power Equation:

Current (A) = Power (W) ÷ Voltage (V) = 20 W ÷ 12 V = **1.67 A.**

Circuit 2: the circuit voltage with two 12 V batteries in series is 24 V:

Current (A) = 20 W ÷ 24 V = **0.83 A.**

Circuit 3: the circuit voltage with four 12 V batteries in series is 48 V:

Current (A) = 20 W ÷ 48 V = **0.42 A.**

With a light bulb of the same power (20 W), if we double the voltage, the circuit current is halved. With a lower current, we can use conductors with a smaller cross-sectional area and keep the voltage drop within the recommended range (the recommended drops for each circuit in an SPV are detailed in Section 8.2 below). For SPV systems we will be working with two broad types of conductors: cables and wires, always made of copper (Figure 51 and Figure 52).

Wire conductors generally lend themselves better to AC circuits, but they are less flexible and therefore more difficult to install. Cable conductors are generally more suitable for DC circuits and for use with high currents. They are more flexible, as they contain a large number of very fine wires, making them easier to install and fix to walls.

Figure 51: Copper wire   Figure 52: Copper cable

Copper cables in which two or three conductors come in the same insulation sleeve (Figure 53). can also be used. For circuits with high currents in an SPV (for example between the batteries and the inverter), it is advisable to use copper cables made for electric welding equipment, designed to carry high currents (Figure 54).

Conductors are classified according to their cross-sectional area and their resistance in Ohms (Ω). Depending on where you are, the cross-sectional area is measured in AWG (American Wire Gauge) with the resistance in Ohms/foot (Table 11), or in mm², with the resistance in Ohms/metre (Table 12). With the AWG system, the cross-sectional area increases as the AWG number decreases:

Figure 53: Cable with three conductors in the same sleeve

Figure 54: Copper cable for welding

AWG Number	Cross-sectional area mm²	Resistance factor, K (Ω/ft)
14	2.00	0.002525
12	3.31	0.001588
10	6.68	0.000999
8	8.37	0.000628
6	13.30	0.000395
4	21.15	0.000249
2	33.62	0.000157
1	42.41	0.000127
0	53.50	0.000099

Table 11: Cross-sectional area of copper cables in AWG and resistance in Ω/ft

Cross-sectional area mm²	Resistance factor, K (Ω/m)
2.5	0.0074
4.0	0.0046
6.0	0.0031
10.0	0.0018
16.0	0.0012
25.0	0.00073
35.0	0.00049

Table 12: Cross-sectional area in mm² and resistance in Ω/m for copper cables

The wires in an installation are differentiated by their colour. In most cases, in direct current, the positive wire is brown or red. The negative (-) is grey, blue or black. Earth wires are almost always green and yellow. AC wires are coloured differently - consult your national regulations to find out what the required colours are. If you can't get hold of wires of the correct colour, as specified in the regulations, use coloured tape and labels to indicate polarity, so that anyone who carries out maintenance work in the future will know which is which.

Conductors installed outside need to be special UV-resistant conductors (Figure 55), or must be protected inside UV-resistant tubing (IEC 60811). Wires that are buried underground need to be housed inside protective tubing.

When it comes to connecting and installing cabling, work in an orderly way and avoid a spider's web of cables, so that identifying problems and doing maintenance is easier.

Figure 55: Installing cabling outside for an SPV

8.2 VOLTAGE DROP

Understanding the voltage drop due to the resistance of a wire is important for SPV systems, especially with very low voltage DC circuits (12V or 24V), as the impact of a 1 V drop on 12 V is proportionally much greater than a 1 V drop in 220 V. If the conductors are under-sized, the voltage drop will result in large losses: the batteries won't charge properly and the load performance will be affected, possibly damaging appliances and/or causing electrical fires. If the wires are over-sized, you'll be spending more than you need to.

The voltage drop in cabling is expressed as a percentage. Table 13 shows the maximum voltage drop for the different circuits of an SPV system:

Circuit	Maximum voltage drop (%)
Module - controller	< 3 %
Controller – battery	< 1 %
Controller – DC loads	< 5 %
Battery - inverter	< 1 %
Inverter – AC loads	< 5 %

Table 13: Rule of thumb for maximum voltage drops

Voltage drop tables are an easy way to work out the maximum distance for a wire with different currents. The table below shows the maximum wire length in meters for 12 V systems, for a voltage drop ≤ 5%:

Cross-sectional area mm²	Current (A)								
	1 A	2 A	3 A	4 A	5 A	6 A	8 A	10 A	14 A
1.5	22m	11m	7m	6m	4m	4m	3m	2m	2m
2.5	38m	19m	13m	9m	8m	6m	5m	4m	3m
4.0	60m	30m	20m	15m	12m	10m	8m	6m	4m
6.0	88m	44m	29m	22m	18m	15m	11m	9m	6m
10.0	150m	75m	50m	38m	30m	25m	19m	15m	11m

Table 14: Maximum wire length in a 12 V system, for a maximum 5% voltage drop
[source: Hankins]

For a 24 V system, multiply the distances in the table by two. For 48 V systems, multiply by four.

8.3 TERMINALS

Terminals are used to ensure a good connection between wires and other components of the system, ensuring minimum losses and no risk of short-circuits. Use terminal blocks or terminal strips to connect several cables together (Figure 56). Use junction boxes at the points where several wires are connected together (Figure 57) and seal them with silicon afterwards so that no insects get in.

Figure 56: Terminal blocks or strips

Figure 57: Junction box

Use pre-insulated brass terminals (Figure 58) and compression pliers (Figure 59) to connect the wires to the different system components.

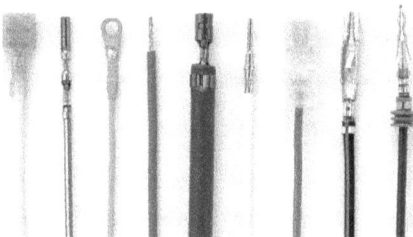

Figure 58: Pre-insulated brass terminals

Figure 59: Compression pliers

Worked Example 12: Voltage drop in cabling

I want to calculate the voltage drop between the module and the charge controller. The length of cable between the components is 5 m, and the maximum module current is 20 A. I want to use a cable with a diameter of 16 mm^2 and a maximum voltage drop of 3 %.

Step 1: Calculate the total resistance of the circuit in Ohms

Work out the total distance of the circuit (there and back): 2 x 5 m = **10 m**.
Work out the resistance factor (K) of the cable using the tables above: **0.0012 Ω/m**
The calculation to find out the total resistance in Ohms is:
Resistance (Ω) = Resistance factor K (Ω/m) x distance (m)
Resistance (Ω) = 0.0012 Ω/m x 10 m = **0.012 Ω**

Step 2: Calculate the voltage drop in Volts

Determine the maximum current between the module and the controller: **20 A**.
The calculation to find out the voltage drop in Volts is:
V_{drop} = Maximum current (A) x Total resistance (Ω)
V_{drop} = 20 x 0.012 = **0.24 V.**

Step 3: Calculate the voltage drop as a percentage

Work out the circuit voltage: **12 V**
The calculation to find out the voltage drop as a percentage is:
V_{drop} (%) = (Voltage drop (V) ÷ Circuit voltage) x 100
V_{drop} (%) = (0.24 ÷ 12) x 100 = **2 %**.

Length of wire (m)	Maximum current (A)	Resistance factor, K (Ω/m)	Total resistance (Ω)	Voltage drop (V)	Voltage drop (%)
10 m	20 A	0.0012 Ω/m	0.012 Ω	0.24	2 %

Perfect! It's less than the maximum voltage drop of 3% between the module and controller. We need to check the maximum cable diameter that the controller terminal can take, to make sure that a 16 mm^2 cable will fit.

9 CIRCUIT PROTECTION

Circuit protection has a dual function: it protects people in case of malfunction and protects system components and appliances. This chapter includes a basic guide. If you are not an electrician, consult a qualified technician and a local supplier of circuit protection components for SPV systems. Read the manuals of the different components and follow the manufacturer's recommendations.

Figure 60: Protection in DC and AC circuits

9.1 FUSES

Fuses in an SPV system are generally either wire fuses or thermomagnetic circuit breakers. Their function is to open a circuit and stop the energy flow when the current exceeds the maximum limit, protecting appliances and users. A fuse consists of a fine wire that burns when the current reaches the level defined by the manufacturer (Figure 62, Figure 63). A thermo-magnetic breaker consists of a switch that opens when the strength of the current reaches the defined value (Figure 61).

Automobile fuses (Figure 63) should not be used in SPV systems that have more than one 12 V battery. If this is the case, you need to use cartridge fuses specifically made for direct current in SPV systems (as in Figure 62). Consult an expert or a SPV supplier. Only use fuses specifically for direct current in a DC circuit, and AC fuses for AC circuits, with the correct voltage.

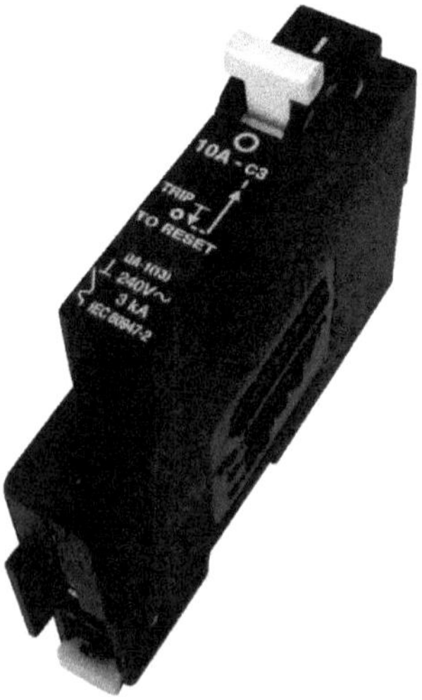

Figure 61: Thermomagnetic circuit breaker    Figure 62: Cartridge fuse

Figure 63: Automobile fuse

All the main circuits in an SPV need to have some form of protection. The battery fuse must be installed on the positive wire, as close as possible to the battery's positive terminal.

For sizing and installing protective elements in an SPV, consult a qualified electrician if need be. Read and follow the manufacturer's recommendations.

Worked Example 13: Sizing fuses

If I have a 12V SPV with two 20 W light bulbs and a 40 W radio, what size fuse do I need to install between the batteries and the charge controller?

Step 1: Calculate the maximum system current

The power of the appliances is: 20 W + 20 W + 40 W = **80 W**
The current is calculated using the Power Equation:
Current (A) = Power (W) ÷ Voltage (V) = 80 W ÷ 12 V = **6.67 A**

Step 2: Calculate the corrected current between the batteries and the controller

Use a safety margin of 20% for the result of the calculation in Step 1.
6.67 A x 1.2 = **8 A**

An **8 A** fuse needs to be installed between the batteries and the charge controller.

9.2 EARTHING

Earthing is a safety measure in an SPV installation, and consists of making an electrical connection from the metallic parts of the equipment to the earth, and an electrical connection from an active conductor to the earth. It performs the following functions:

➢ **Protection against lightning:** it protects the users and appliances in case of a lighting strike during an electric storm.

➢ **Protection in case of short-circuits:** if a piece of equipment short-circuits (for example, when the insulation on a wire gets worn or a terminal comes loose and makes contact with another cable or the metal case of a component), it prevents the metallic parts of the equipment from being "live", protecting the end-users.

This consists of a copper spike buried underground, connected to a wire with a given cross-sectional area. This wire connects to the conductive (metallic) parts of the system, or to specific parts of an appliance or circuit (for example, to the earth terminal on an inverter). The cross-sectional area of earth cables is calculated as a function of the diameter of the phase conductors in the system (Table 15).

Diameter of phase conductors (S_f) in mm²	Minimum diameter of earth cable (S_t) in mm²
$S_f \leq 16$ mm²	$S_t = S_f$
$S_f < 16$ mm² ≤ 35 mm²	$S_f = 16$ mm²

Table 15: Calculating the cross-sectional area of the earth wire [source: Pareja Aparicio]

Consult the earthing regulations in your country, and read the component manuals, following manufacturers' recommendations.

Figure 64: Burying a copper rod for an earth connection

10 SIZING AN SPV SYSTEM

There are many different ways of sizing a low-power SPV system, ranging from simple spread sheet calculations, to sophisticated simulation programs. The following presents a simplified method for sizing low power SPV systems. For larger systems, consult an expert and the technical resources in Appendix 16.1. Sizing an SPV system consists of calculating energy demand and production, and sizing the system components accordingly.

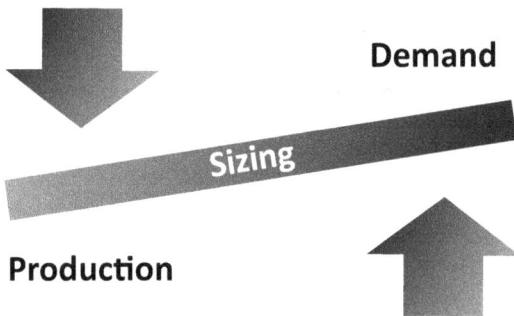

Figure 65: System sizing, balancing generation with demand

If your system has excess capacity (in other words, if it produces much more energy than what is consumed), it'll be ecologically and economically inefficient. If it's undersized, it won't be able to supply the energy required by end-users and there'll be recurring problems with the batteries, shortening their life. This chapter will help you do the basic calculations for a low-power system, to find a solution that meets the available budget.

The sizing method presented here consists of a series of steps using six calculation sheets. It is viable for low power SPV systems of ≤ 500 Wp. Amp-hours (a unit of charge) are used instead of watt-hours (a unit of energy) for the calculations, in order to take into account a more realistic module and battery performance, depending on their temperature and voltage. This method is appropriate for systems where the charge controller is not MPPT (maximum power point tracking). The calculation sheets can be copied from Appendix 16.2 and filled out by hand, or they are available to download in Excel format from the following web address: www.itacanet.org.uk/doc-archive-eng/solar/solar_sizing_spreadsheet_08-13.xlsx.

There is no quick fix for sizing a system. Repeat the sizing process several times with different assumptions and compare the results and costs. Take the final design to an expert in your area and ask for their feedback. Table 16 presents a summary of the content in each step and the corresponding calculation sheet:

#	Calculation sheet	Description
1	Solar resource	Calculating the solar resource for the site
2	Demand	Estimating the energy demand of the system
3	Modules	Sizing the photovoltaic module
4	Batteries	Sizing the battery bank
5	Controller and/or inverter	Sizing the charge controller and inverter
6	Cabling & fuses	Sizing wires and fuses

Table 16: Steps for sizing a low power SPV

The initial investment for an SPV system is relatively high. If first time you size and cost the system, you run over the available budget, re-size and look for a lower cost solution. Systems can always be expanded in the future. Alternatively, think about incorporating another source of energy generation. Above all, look for the simplest solution!

Below are the basic steps for sizing a low power system, based on a case study, for a community centre in Quetzaltenango, Guatemala. The centre has just been built, in a town where there is still no grid connection. The closest power lines are 8 km away. The national power company has stated that the extension of the grid to this community "is planned" within the next 20 years. The Directors were initially thinking about using a diesel plant to provide electricity to the centre. However, due to the increase in fuel prices during the construction of the centre, they want to see whether an SPV system would be a viable solution, with lower operating costs.

10.1 SITE VISIT

Before sizing a system, the first step is to do a site visit, where the installation is planned. The site visit will provide the information you need to size and cost the system: make sketches of the building and its surrounding, and take photos. Figure 66 shows an example, of the case study in question, with a building plan, including the room layout, dimensions, and details of the surroundings.

To maximise the energy collected by the generator and avoid shadows cast by the neighbouring building and trees to the south-west and west of the building, you've agreed

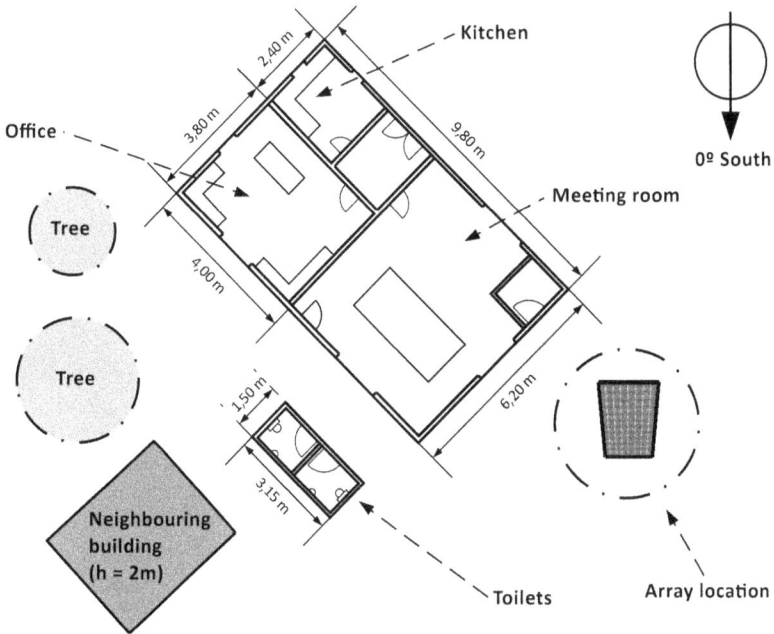

Figure 66: Plans of community centre

with the Directors that the optimal location for the modules is to the south-east, close to the small room in the east corner of the building, suitable for housing the battery bank. The room has air vents and sufficient storage space. The controller, inverter and electrical distribution board can be mounted on the other side of the wall, in the meeting room.

It has been envisaged that the system will have DC lights and a small inverter to connect to the AC cabling in the building, supplying energy to laptop computers, a printer and an LCD television. These are appliances that the centre already possesses, and are relatively efficient. The following list presents the most important questions for the site visit:

> **Existing energy sources:** what other energy sources are there at the site? How far away is the closest connection to the national electricity grid? Are there diesel or petrol generators?

> **Service and repair centres available in the area:** is there specialised labour in the area that can repair SPV equipment?

> Is there local labour available that can install and maintain the system, and what is the going rate/hour for worke?

> **Location of the photovoltaic generator:** is there somewhere with no shading where the generator can be placed, at the optimal orientation and inclination? Where is geographical South?

➢ **Location of the controller/ batteries/inverter:** is there a suitable place for the controller/batteries/inverter? Will you need to build a new outbuilding, or can they be housed in the existing building?

➢ **Energy demand:** What are the load appliances? What is their power in watts or amps? For how many hours each day will they be in use?

➢ **Measuring distances and quantities:** what are the dimensions of the building to be electrified? How far away will the photovoltaic generator be from the controller / batteries/inverter? Where in the building will you need to install lights, fuses, and plugs?

If the site is in a remote place where access is difficult, remember that it can be expensive to go back for a measurement that you have forgotten. Spend sufficient time gathering data – it will make it easier to size and cost the system.

10.2 SOLAR RESOURCE

To calculate the solar resource at our site, we use climate data of solar irradiation in kWh/m^2/day, or Solar Peak Hours, as in Chapter 2. Try to find solar irradiation data from the nearest meteorological station to your site. The free program RETScreen contains a database of solar irradiation figures for thousands of places around the world (see Appendix 16.1). Appendix 16.5 contains data for a large number of places in Latin America.

There are two methods for calculating the solar resource: the *critical month* method and the *average* method. The first takes the minimum level of solar irradiation over the course of the whole year. The second takes the average over the whole year. It is always advisable to use the critical month method (the minimum over the whole year) to size a system: that way you ensure that the system is capable of generating sufficient energy during months when there's less sun, and for successive cloudy days. The batteries will also last longer this way.

The critical month method raises the cost of the system and it may be that for part of the year it will produce more energy than is consumed (excess capacity), especially at latitudes further north or south than the tropics. Make a decision based on the following criteria:

➢ **The variation in solar irradiation over the year:** in tropical regions the difference between the critical month and the average is between 15 – 25%. The further north or south of the tropics, the greater this difference is. This means that for a small, non-critical installation in the tropics, using the average method is feasible. For critical installations outside the tropics, use the critical month method to ensure sufficient energy production in the winter months.

➢ **Budget and type of installation:** for a small house where only lighting is required, the budget will be limited and end-users will generally be able to adapt more easily to a lack of energy during cloudy days. In these cases, the average method can be used. On the other hand, for a rural clinic, the budget will be larger, the installation is more critical, and a reliable and constant energy supply will be more important throughout the year. In this case, use the critical month method.

For the community centre in Quetzaltenango, we have figures for the average solar radiation per day on a horizontal surface, for each month of the year, in kWh/m²/day. Figure 67 below shows Spread Sheet 1 – Solar Resource, which allows us to work out the Peak Sun Hours for the system design.

The steps to follow are as follows:

Firstly, enter the installation details, the client, your name, the date the study was carried out, the system location and coordinates. Remember that all documentation you produce for the project must be well presented and intelligible for anyone who works on the system in the future.

SPREAD SHEET	1 SOLAR RESOURCE

Prepared by	Community Centre
Prepared for	OS
Date	31/02/2012
Location	Queztaltenango, Guatemala
Coordenates	14,8 N / 91,5 W

NOTE:

Enter data in green cells

Results are shown in yellow cells (remove spread sheet protection to make changes in cells other than green)

Step	Solar irradiation on a horizontal surface (kWh/m2/día) - Peak Sun Hours												
	January	February	March	April	May	June	July	August	September	October	November	December	*Average*
1	5.31	5.76	6.07	6.03	5.55	5.31	5.55	5.28	4.70	4.80	5.00	5.12	*5.37*

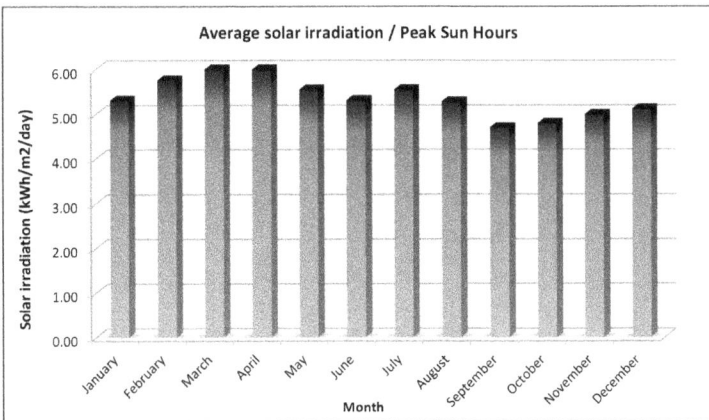

2	Critical month	4.7

Figure 67: Spread sheet 1 - solar resource

STEP 1:

There is a space to fill in the irradiation or Peak Sun Hours below each month. Enter the relevant figures. The annual average is calculated in the "Average" cell (the calculation is the sum of all the months, divided by 12): 5.37 PSH.

STEP 2:

The last cell on the sheet is used to calculate the critical month, which is the month with the least irradiation in the whole year: 4.7 PSH.

10.3 DEMAND CALCULATION

We have calculated how much energy we will get from the sun throughout the year and opted to use the critical month method, using 4.7 PSH for the system specification. Now we have to calculate how much energy will be consumed in the centre every day of the week. The demand calculation is an estimation based on a list of all appliances that are used daily, their power in watts, and how many hours each day they will be in use.

Worked Example 14: Energy demand

If I have two 15 W low energy light bulbs and I use them for 4 hours per day (both of them), how much energy have I consumed in Wh?

Quantity	X	Power (W)	X	Hour of use/day (h)	=	Consumption(Wh/day)
2	X	15	X	4	=	120 Wh/ day

Energy consumed = **120 Wh/day**

In order to calculate the total system demand in amp-hours, we need to determine the system voltage. As we saw in Chapter 8, high currents require thicker wires, in order to avoid excessive voltage drops. This is why, for more powerful SPV systems where the current between modules, controller, batteries and inverter is higher and the distances are greater, voltages of 24 V or 48 V tend to be used. This reduces the current and means that smaller wires can be used, reducing costs.

For low power SPV systems, it is advisable to choose a system voltage of 12 V, as currents tend to be lower and you will be able to use thinner cables. A 12 V system allows you to use DC light bulbs and efficient 12 V DC appliances. If you specify an inverter, you need to define the AC output voltage and frequency. This will be determined by the voltage and frequency of the national grid in your country.

Having gathered data from the community centre in Quetzaltenango and made an analysis of efficient DC light bulbs available on the local market, we have decided that compact fluorescent light bulbs are the most suitable option. In Figure 68 you can see the Spread Sheet 2 – Energy Demand.

To carry out the calculations, follow the steps below:

STEP 1: CALCULATE THE LOADS

Fill in the "Load" column with the names of DC and AC appliances and where they will be used. In the "Quantity" column, indicate the quantity of each appliance. In the "Power (W)" column,

Step	Load	Quantity	x	Power (W)	x	Hours of use/day (h)	=	DC demand (Wh/day)	o	AC demand (Wh/day)
	SPREAD SHEET — 2 ENERGY DEMAND									
	Loads in Direct Current (DC)									
	Lighting kitchen	1	x	11	x	2	=	22		
	Lighting meeting room	4	x	15	x	3	=	150		
	Lighting bathroom	2	x	11	x	1	=	22		
1	Lighting office	2	x	15	x	4	=	120		
	Loads in Alternating Current (AC)									
	Laptop computer	2	x	45	x	3			=	270
	Mobile phone charger	2	x	4	x	2			=	16
	Printer	1	x	35	x	1			=	18
	Television	1	x	90	x	1			=	90
2	*Daily demand in DC (Wh)*							314		
	Daily demand in AC (Wh)									394
3	*Percentage of losses in DC (%)*							20%		
	Percentage of losses in AC (%)									35%
4	*Losses in DC (Wh) [daily demand x losses]*							63		
	Losses in AC (Wh) [daily demand x losses]									138
5	**Corrected demand in DC (Wh/day)** *[daily demand x losses]*							377		
	Corrected demand in AC (Wh/day) *[daily demand x losses]*									531
6	**Total daily demand (Wh/day)** *[corrected DC demand + corrected AC demand]*							908		
7	**System voltage (V)**							12		
8	**Required daily charge(Ah)** *[total daily demand ÷ system voltage]*							76		

Figure 68: Spread Sheet 2 – Energy Demand

write its power in watts. In the "Hours of use/day" column, indicate how many hours per day it will be in use.

Fill in the "Load" column with the names of DC and AC appliances and where they will be used. In the "Quantity" column, indicate the quantity of each appliance. In the "Power (W)" column, write its power in watts. In the "Hours of use/day" column, indicate how many hours per day it will be in use.

STEP 2: CALCULATING DAILY DEMAND IN AC AND DC

Calculate the total energy required per day from all DC appliances:

 22 + 150 + 22 +120 = **314 Wh/day**

 And AC:

 270 + 16 + 18 + 90 = **394 Wh/day**

STEP 3: ESTIMATING THE DC AND AC LOSSES

We have to estimate the losses from system inefficiencies. For low power SPV systems, we can estimate there will be DC losses of 20% and AC losses of 35%. Enter the losses in the corresponding boxes.

STEP 4: CALCULATING THE DC AND AC LOSSES

Calculate the actual losses by multiplying the daily demand by the percentage losses, in DC and AC:

 314 Wh/day x 0.2 = **63 Wh/ day** (DC)
 394 Wh/day x 0.35 = **138 Wh/day** (AC)

STEP 5: CALCULATING THE CORRECTED DEMAND IN DC AND AC

Add the losses to the daily AC and DC demand, to calculate the corrected demand:

 314 + 63 = **377 Wh/day** (DC)
 394 + 138 = **531 Wh/day** (AC)

STEP 6: CALCULATING THE TOTAL CORRECTED DEMAND

Calculate the total corrected daily demand by finding the total of the AC and DC corrected demand:

 377 + 531 = **908 Wh/day**

STEP 7: SYSTEM VOLTAGE

Enter the system voltage: **12 V**

STEP 8: CALCULATING THE SYSTEM'S DAILY LOAD

Calculate the total daily load of the system by dividing the total corrected demand by the system voltage:

908 Wh/day ÷ 12 V = **76 Ah**

Congratulations! You have calculated the energy demand for the system.

Sometimes the power of an appliance is rated in amps, not watts: you can use a simple calculation to work out its power in watts, using the Power Equation:

Worked Example 15: Energy demand

I have a mobile phone charger that consumes 0.1 A at 110 V. What is its power consumption in (W)?

Power (W) = 0.10 A x 110 V = **11 W**

Once you've done a first demand calculation, look at how you can improve the system's efficiency and reduce costs. Figure 69 shows an analysis of consumption by category, including the DC losses (20%) and AC losses (35%):

You can see that the consumption by computers (40%) and the television (13%) make up a major part of the demand (53% of the total), largely due to inverter losses in the AC circuits. The DC lighting represents a total of 41% of the demand.

Losses in DC circuits are 7% of the total. On the other hand, losses in AC circuits are more than twice that: 15%. This is why it is better to use efficient DC appliances instead of AC, and avoid inverter losses.

If your demand calculation is excessive, the cost of the system will shoot up and there will be excess capacity. If you underestimate the system demand, there will be recurring problems with the batteries and the operating costs will increase. Make a careful and realistic estimation and check it through with the end users.

DEMAND ANALYSIS

		Demand (+ losses)	% of total
DC	Lighting kitchen	26	3%
	Lighting meeting room	180	20%
	Lighting toilet	26	3%
	Lighting office	144	16%
AC	Laptop computer	365	40%
	Mobile phone charger	22	2%
	Printer	24	3%
	Television	122	13%
	TOTAL	908	100%

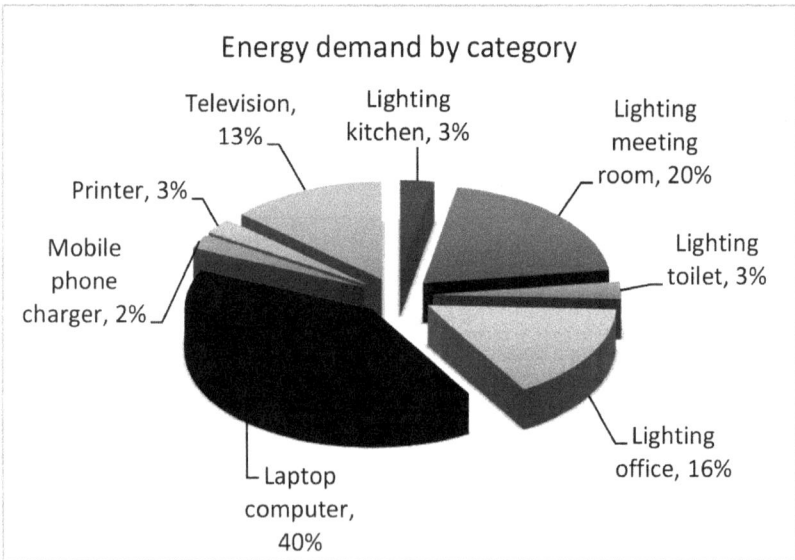

Figure 69: Analysis of energy demand by category

Step	Daily charge (Ah)	÷	PSH critical month	=	System charging current (A)
1	76	÷	4.7	=	**16**

	Module	Kyocera KC 80	
	Nominal voltage (V)	12	
2	Maximum power (W)	80	
	Maximum current Isc (A)	4.97	
	Current SOC (A)	4.00	

3	**Number of modules**	4	

Figure 70: Spread Sheet 3 – Modules

10.4 Modules

We have calculated the solar resource for our site in Peak Hours (**4.7 PSH**) and the daily energy requirement (**76 Ah**). Now we can size the modules. After comparing the prices and performance of various 12 V modules available locally, we opt for the module we saw in Chapter 2: a Kyocera KC80. Based on the manufacturer's data sheet and the I-V curve under Standard Operating Conditions (SOC), we calculate that the module has an output current of 4 A (see Figure 23).

Table 17 shows the calculation which determines the load current of the system:

Calculating the load current of the system					
Daily system load (Ah)	÷	PSH critical month	=	**System load current (A)**	

Table 17: Calculation to find the system load current

Figure 70 shows Spread Sheet 3 – Modules.

Calculation to find the system load current:

STEP 1: CALCULATING THE SYSTEM LOAD (A)

In the "Daily system load" cell enter the daily load that we calculated in Sheet 2: 76 Ah. In the "PSH critical month" cell, enter the Peak sun hours from Sheet 1 – Solar Resource: **4.7**.

The peak hours in the critical month divided by the daily load on the system gives us the load current of the system in amps: **16 A**.

STEP 2: DATA OF THE SELECTED MODULE

Enter the data of the module we have chosen; its model (Kyocera KC80), voltage (12 V), maximum power (80 W) and current under Standard Operating Conditions[SOC] (4 A).

STEP 3: CALCULATING THE NUMBER OF MODULES REQUIRED

To work out how many modules we need, the calculation is as follows:

Calculating the number of modules				
System load current (A)	÷	Module current (SOC)	=	**Number of modules**

Table 18: Calculation to work out the number of modules required

> *If, for example, the result of the module calculation comes out as a 3.2, you need to round up to 4 modules.*

The number of modules = 16 ÷ 4 = **4 modules**.

Once you have calculated the number of modules, you need to make a schematic diagram showing the connections. In this case, as the voltage of each Kyocera module is 12 V and the system voltage is 12 V, the connection is made in parallel, so the voltage will remain the same and the current adds up (Figure 71 Parallel, series and mixed connections are explained in more detail in Chapter 12.

Good work! We have sized the module array and drawn up the electrical schematic. On to the next step...

10.5 BATTERY BANK

For the community centre in Quetzaltenango, we've researched the batteries available on the local market, comparing price per cycle (see Chapter 4), and decided on a Trojan T-105.

To size the battery bank, four parameters that need to be defined:

1. **Daily system load in Ah:** this is the result from Spread Sheet 2 – Energy Demand: 76 Ah.

2. **Maximum depth of discharge:** this depends on the battery you have chosen. Remember: in general, starter batteries should have a depth of discharge of less than 15%; hybrids should be less than 35%, while deep cycle batteries can reach depths of 50%. With the Trojan T-105 traction battery, the DoD can reach a maximum of 50%.

PHOTOVOLTAIC MODULES

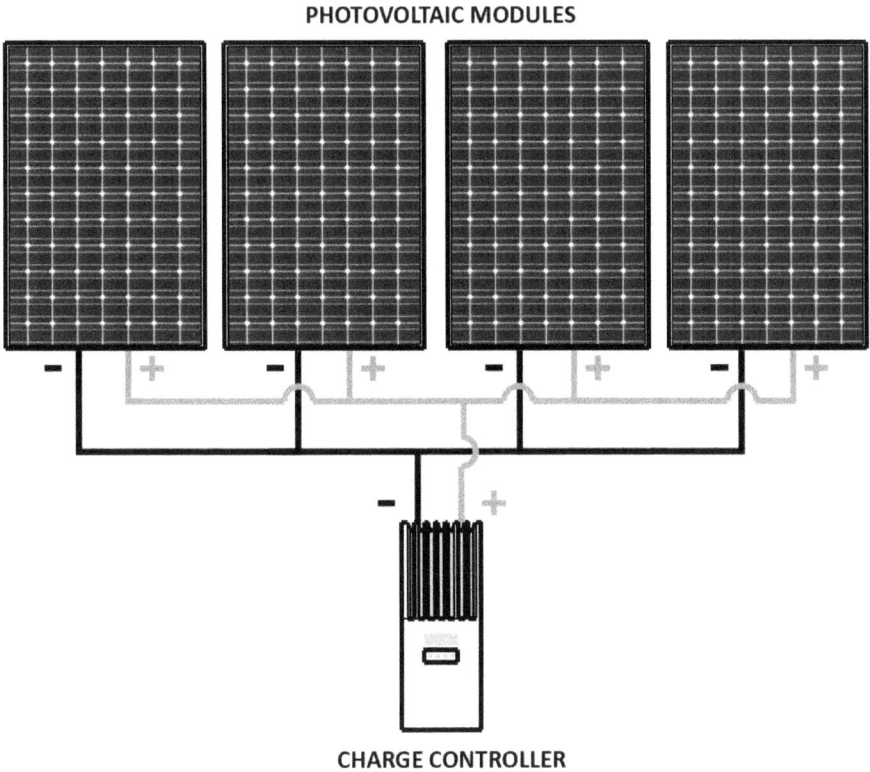

CHARGE CONTROLLER

Figure 71: Four 12 V modules connected in parallel for a 12 V SPV

3. **Days of autonomy:** this refers to how many days of reserve energy are required, to provide power over a succession of cloudy days. For low power SPV systems, between 1 and 4 days is recommended.

4. **Battery capacity at C20:** check the manufacturer's data sheet, and look for the battery's nominal capacity at C20. For the Trojan T-105, this is 225 Ah. Watch out – many manufacturers give their batteries' specification at C 100, which is always higher than the capacity at C20. For the same battery, the C100 capacity is 250 Ah.

Table 19 shows the calculation for the battery bank capacity:

Calculation to specify the battery bank capacity in Ah					
Daily system load (Ah)	×	Days autonomy	÷	Days depth of dischargea	= Battery bank capacit (Ah)

Table 19: Calculation for sizing the battery bank capacity in Ah

SPREAD SHEET   4 BATTERIES

Step	Daily charge (Ah)	x	Days of autonomy	÷	Depth of discharge*	=	Battery bank capacity (Ah)
1	76	x	3	÷	50%	=	454

2	Battery	Trojan T-105
	Nominal voltage (V)	6
	Capacity @ C20 (A)	225

3	Batteries in series	2
	Batteries in parallel	2
	Number of batteries	4
	Total capacity (Ah)	450

Equal: yes or no?

4	Useful life @ selected depth of discharge* (cycles)	1500
	Estimated useful battery life (life)	4

Figure 72: Spread Sheet 4 – Battery bank

Now we move on to Spread Sheet 4 – Battery bank:

The steps for sizing the battery bank are:

STEP 1: CALCULATE THE CAPACITY OF THE BATTERY BANK

Enter the daily load in Ah in the corresponding cell. This was calculated in Spread Sheet 2 – Demand:

76 Ah.

Enter the days of autonomy required (between 1 and 4 days): **3 days.**
Enter the maximum depth of discharge for the battery you have chosen: **50 %.**
The calculation for the capacity required for the battery bank is:

$76 \times 3 \div 0.5 = $ **454 Ah**

STEP 2: ENTER THE BATTERY DATA

Enter the make and model of the battery (Trojan T-105), its nominal voltage (6 V) and its nominal capacity at C20:

225 Ah.

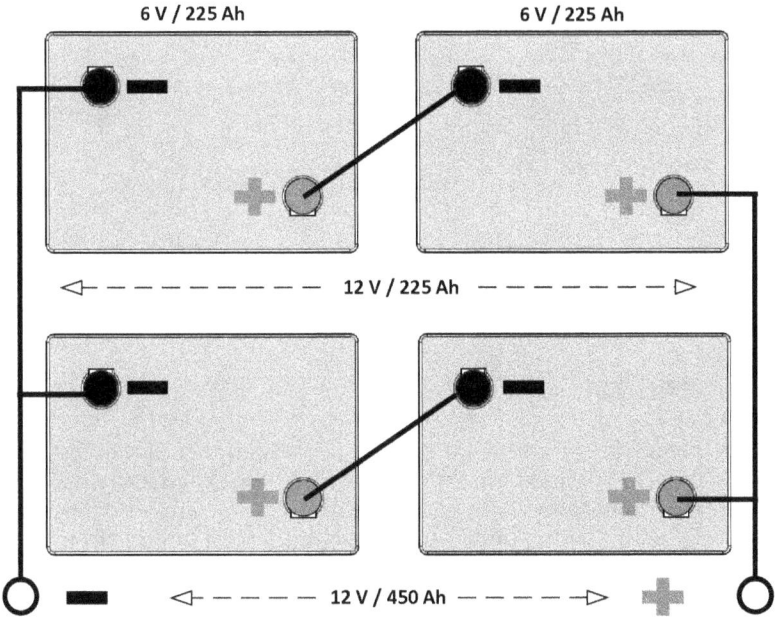

Figure 73: Series-parallel connection for the battery bank

CALCULATING THE NUMBER OF BATTERIES IN SERIES, PARALLEL AND IN TOTAL

In this case we have selected a battery with a nominal voltage of 6 V. However, our system is 12 V. In order to use these batteries in our system, we need to have two batteries connected in series to get a voltage of 12 V. With connections in series, the voltage adds up and the current stays the same. With connections in parallel, the voltage stays the same and the current adds up. We need to get to 12 V with a total capacity of 453 Ah, calculated in Step 1.

Two batteries connected in series in the first row give 12 V and 225 Ah. Taking the two rows in series and then connecting them in parallel, we get to 12 V with 450 Ah: almost equal to the capacity calculated in Step 1. Enter the number of batteries in series and parallel in the spread sheet, and add these numbers to get the total quantity. 4 batteries Figure 73 shows the connections schematic.

STEP 4: ESTIMATING USEFUL BATTERY CYCLE LIFE

The last two cells on the sheet show a calculation to estimate the cycle life of the battery bank. From the battery sheet we know that at a 50% depth of discharge, the cycle life is approximately 1500 cycles. Enter this figure in the corresponding box in the spread sheet. Table 20 shows the calculation for estimating the cycle life in years:

Calculation to estimate the useful life of a battery				
Useful life at 50% DoD (cycles)	÷	Number of days in a year	=	Cycle life in years
1500	÷	365	=	4 Years

Table 20: Calculating the cycle life of a battery

This information needs to be given to the system operators, so that they can plan to buy new batteries after approximately four years.

10.6 CONTROLLER AND INVERTER

Nearly there! We just need to size the controller and inverter. For our system, we are using DC lighting and AC for the equipment in the office. To size the controller, we need to calculate the maximum short-circuit current of the modules. We calculate this by multiplying the maximum short-circuit current of each module (Isc) by the total number of modules, adding a 25% safety margin.

We also have to calculate the maximum output current to the DC load. We add up the power in watts of all the DC appliances (from Spread Sheet 1 – Demand), and divide by the system voltage to convert to amps. We add a safety margin of 35%. Our controller must have a nominal current that is higher than either of these two figures.

To size the inverter, we have to add up the power in watts of all the AC appliances, and add a 25% safety margin. The nominal power of the inverter must be greater than this result. Let's look at Spread Sheet 5 – Controller and Inverter:

CONTROLLER:

STEP 1: CALCULATE THE CORRECTED MAXIMUM CURRENT FROM THE MODULES

From Sheet 3 – Modules, take the maximum short circuit current of one module and multiply by the total number of modules in series:

4.97 A x 4 = 19.7 A. Round up to **20 A.**

Multiply the maximum current of the generator by the safety margin of 25%, to get the corrected maximum current of the generator in amps:

20 × 1.25 = **25 A.**

SPREAD SHEET | 5 CONTROLLER & INVERTER

CONTROLLER

Step	Max. current modules (A)	x	Safety margin (%)	=	Corrected maximum current modules (A)
1	20	x	25%	=	25.0

	Max. DC load current (A)	x	Safety margin (%)	=	Corrected maximum current DC load (A)
2	10.25	x	25%	=	12.8

Which larger?

3	**Maximum current charge controller (A)**	**25**

INVERTER

	Max. power AC load (W)	x	Safety margin (%)	=	Max. corrected power AC load (W)
4	223	x	25%	=	**278.8**

5	**Minimum nominal power inverter (W)**	**300**

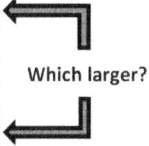

Figure 74: Spread Sheet 5 – Controller and Inverter

STEP 2: CALCULATE THE CORRECTED MAXIMUM LOAD CURRENT

From Spread Sheet 2 – Demand, we add up the power in watts of all the DC appliances that could be turned on at the same time, and divide by the system voltage to get the result in amps:

Kitchen lighting: 1 × 11 W = 11 W
Meeting room lighting: 4 × 15 W = 60 W
Bathroom lighting: 2 × 11 W = 22 W
Office lighting: 2 × 15 W = 30 W

TOTAL = 123 W

123 W ÷ 12 V = **10.25 A.**

Multiply the maximum load current by the safety margin of 25% to get the corrected maximum load current in amps:

10.25 × 1.25 = **12.8 A.**

STEP 3: CALCULATING THE MAXIMUM CURRENT OF THE CONTROLLER

We choose the higher figure from the results of the calculations in Step 1 and Step 2: **25 A**.

Our charge controller needs to have a minimum nominal current rating of **25 A.**

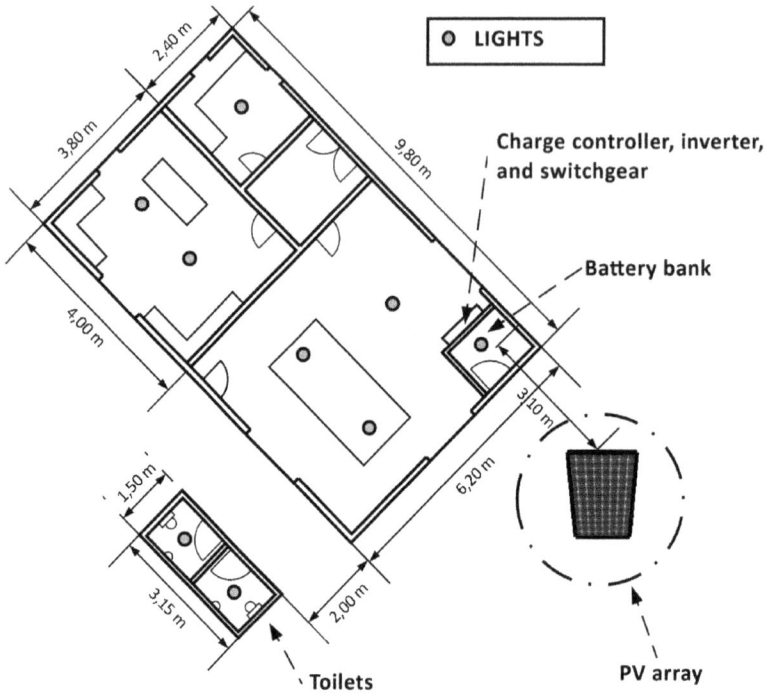

Figure 75: Lighting layout

INVERTER:

STEP 4: CALCULATING THE MAXIMUM POWER OF AC LOADS

From Spread Sheet 2 – Demand, we add up the power in watts of all the AC appliances:

Laptop computer: 2 × 45 W = 90 W
Mobile phone charger: 2 × 4 W = 8 W
Printer: 1 × 35 W = 35 W
Television: 1 × 90 W = 90 W

TOTAL = 223 W

This is the maximum power in the AC circuit with all the appliances switched on at the same time.

We multiply the maximum power of the AC load by the 25% safety margin, to calculate the corrected maximum power of the load in watts:

223 × 1.25 = **278.75 W.**

STEP 5: WORKING OUT THE MINIMUM NOMINAL POWER OF THE INVERTER

The result of the calculation in the previous step shows that the maximum corrected power of the AC load is 255 W. We round this up to work out the minimum nominal power rating of the inverter:

300 W

10.7 WIRING AND FUSES

Cable sizing involves calculating the required cross-sectional area of a cable in a circuit for the current it will carry: if the cable is too thin, there'll be an excessive voltage drop and energy loss in the system; if it's too thick, installation is more complicated and you'll be wasting money.

After our site visit to the community centre and data gathering, we've drawn up a plan of the building with dimensions. The plan helps us to calculate the length of cable runs in each circuit. In this example, we will be sizing cables and fuses for the DC network. Figure 75 shows the plan of where the fluorescent light bulbs will be installed in each room.

Remember that to calculate the length of wire in a circuit, you need to add up the distance both ways, including the bends and curves that the wire will have to take in its path around the building. For the final system design, lengths need to be measured accurately. If the length of the wire between the controller and the light bulb in the kitchen is 9.2 m, the length to use for sizing the wires will be 18.4 m.

To work out the maximum current in a circuit, add up the load in Amps of all the appliances that could be switched on at the same time. Following the previous example, in the meeting room there will be four 15 W light bulbs, so the maximum current in the circuit will be 4 x 1.25 A = 5 A.

Once we have measured the lengths of the cables and calculated the maximum currents of the equipment in each circuit (referring to the manufacturers' figures), we can enter the data in Spread Sheet 6 – Wiring and Fuses. Consult Chapter 8 to work out the resistance factor K (in Ohms/m) of different conductors. For this installation, we've established that the voltage drop in all the circuits will be ≤ 2 %:

WIRING:

STEP 1: SIZE THE CABLES FOR A ≤ 2 % VOLTAGE DROP IN THE CIRCUIT

In the "Circuit" column we put a description of the circuit. If the circuit has several appliances, indicate how many and their current (for example: 2 x 1.25 A).

In the "Length of wire (m)" column, enter the total length in metres, adding up the distance both ways.

SPREAD SHEET	6 WIRING AND FUSES

WIRING

Step	Circuit	Cable run length (m)	Max. current (A)	Wire resistance factor K (Ω/m)	Total resistance (Ω)	Voltage drop (V)	Voltage drop (%)
1	Lighting controller > kitchen (1 x 1.25 A)	18.40	1.25	0.0074	0.136	0.170	1.42%
	Lighting controller > meeting room (4 x 1.25 A)	5.00	5.00	0.0074	0.037	0.185	1.54%
	Lighting controller > bathrooms (2 x 0.92 A)	20.30	1.84	0.0046	0.093	0.172	1.43%
	Lighting controller > office (2 x 1.25 A)	22.40	2.50	0.0031	0.069	0.174	1.45%
	Modules > controller	7.15	25.00	0.0012	0.009	0.215	1.79%
	Controller > batteries	2.75	25.00	0.0031	0.009	0.213	1.78%
	Batteries > invertor	2.80	66.67	0.0012	0.003	0.224	1.87%

FUSES

Step	Circuit	Max. power (W)	Max. current (A)	Nominal current fuse rating (A)
2	Controller > batteries	-	25.00	30
		-	-	-

Figure 76: Spread Sheet 6 – Wiring and Fuses

In the "Conductor resistance factor K (Ω/m)", enter the resistance factor of the wire you intend to use in the circuit.

The next columns show the results for the total resistance (Ω) of the circuit, the voltage drop in Volts and as a percentage (%). The result in the last column "Voltage drop (%)" is the most important (see Chapter 8).

Figure 76 shows the calculations for each DC circuit, ensuring that the voltage drop will not exceed 2 %. You can see that the circuits with higher current and resistance are "Generator > controller" and "Batteries > inverter), which require cables with a resistance factor K of 0.0012 Ω /m. These wires have a cross-sectional area of 16 mm^2.

FUSES:

STEP 2: SIZE THE FUSES IN DC CIRCUITS

In the "Circuit" column, enter a description of the circuit. In the example above, we are sizing the fuse for the circuit from the charge controller to the batteries.

In the "Maximum current (A)" column, enter the maximum current that will come from the modules (calculated in Sheet 5):

24.64 A.

Apply a safety margin of 20%. The calculation is as follows:

Fuse capacity (A) = Maximum current (A) x 1.20 = 24.64 A x 1.2 = **30 A.**

Congratulations! You have finished sizing your SPV system! After the first round of calculations, go back over them and look at how the system can be improved, making it more efficient and reducing cost.

Circuit protection is important. Sizing fuses and protections for SPV systems which include inverters must be done by a qualified person, and must always comply with the national codes and regulations.

11 Economic analysis

Economic criteria will ultimately determine whether an installation goes ahead or not. Economic analysis that weighs up the viability of the system compared to other alternatives is useful. There are, of course, other factors to take into account, such as the importation of equipment from abroad, warranties offered by the manufacturers, the technical service a supplier can offer, and the suitability of a given component to the needs and habits of the users.

11.1 Budget

After sizing the system, make a detailed list of all the components and materials needed and get quotes from suppliers in the area. This information will help you prepare the final budget. Get quotes from at least two different suppliers in order to compare prices.

Remember that it is very important to include the exact specifications for each component and transmit this information clearly to the supplier, otherwise you may receive quotes for the cheapest equipment or for components that won't work in the system you've designed. Once you have the quotes, compare and work out the final cost/watt for the system as a whole. Figure 77 shows a preliminary budget and cost analysis for the system that was specified in the previous Chapter. It compares two options and the prices from different suppliers.

As you can see, Option 2 will result in a small saving, with a cost of 18.98 US$/Wp, compared to 19.78 US$/Wp for Option 1. In both cases, the cost of the modules represents approximately 30% of the total cost of the system. The batteries are approximately 15%. An analysis like this makes it easier to take decisions and allows you to quickly visualize the cost percentage that each part of the system represents. The outcome, the final price in US$/Wp, allows you to compare different options and to compare the total system cost with other systems and prices.

BUDGET

Peak system power (Wp)	320

OPTION 1

Component	Manufacturer and model	W / A / Ah	Quantity	Unit price (US$)	Total price (US$)	Price (US$ per W,A,ah)	% of total
PV module	Kyocera KC80	320	4	$ 527.00	$ 2,108.00	$ 1.65	33%
Charge controller	Morningstar Sunsaver DUO 25 A	25	1	$ 263.00	$ 263.00	$ 10.52	4%
Batteries	Trojan T 105 RE 6V 225 A/H	225	4	$ 255.00	$ 1,020.00	$ 1.13	16%
Inverter	Morningstar Sure Sine SI-300-220V	300	1	$ 399.00	$ 399.00	$ 1.33	6%
Panel mount	Galvanized steel	-	1	$ 423.00	$ 423.00	-	7%
Lights	Various	-	15	$ 20.00	$ 300.00	-	5%
Cables	Various	-	1	$ 476.00	$ 476.00	-	8%
Terminals	Various	-	1	$ 99.00	$ 99.00	-	2%
Fuses	Various	-	1	$ 167.00	$ 167.00	-	3%
Labour	-	-	1	$ 450.00	$ 450.00	-	7%
Transport	-	-	1	$ 625.00	$ 625.00	-	10%
							100%

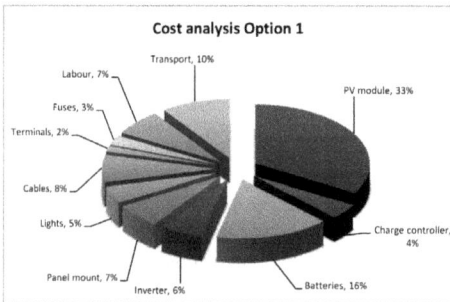

Subtotal (US$)	$ 6,330.00
VAT (18%)	$ 1,139.40
TOTAL	$ 7,469.40
Cost (US$/Wp)	$ 19.78

OPTION 2

Component	Manufacturer and model	W / A / Ah	Quantity	Unit price (US$)	Total price (US$)	Price (US$ per W,A,ah)	% of total
PV module	Solarworld SW80	320	4	$ 445.00	$ 1,780.00	$ 1.39	29%
Charge controller	Steca Solarix PRS 3030 30 A	30	1	$ 207.00	$ 207.00	$ 6.90	3%
Batteries	US Battery U2200 &V 232 A/H	232	4	$ 217.00	$ 868.00	$ 0.94	14%
Inverter	Morningstar Sure Sine SI-300-220V	300	1	$ 382.00	$ 382.00	$ 1.27	6%
Panel mount	Galvanized steel	-	1	$ 646.00	$ 646.00	-	11%
Lights	Various	-	15	$ 20.00	$ 300.00	-	5%
Cables	Various	-	1	$ 476.00	$ 476.00	-	8%
Terminals	Various	-	1	$ 99.00	$ 99.00	-	2%
Fuses	Various	-	1	$ 167.00	$ 167.00	-	3%
Labour	-	-	1	$ 398.00	$ 398.00	-	7%
Transport	-	-	1	$ 749.00	$ 749.00	-	12%
							100%

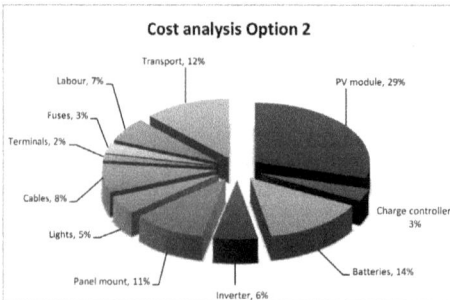

Subtotal (US$)	$ 6,072.00
VAT (18%)	$ 1,092.96
TOTAL	$ 7,164.96
Cost (US$/Wp)	$ 18.98

Figure 77: Preliminary budget and cost analysis

11.2 LIFE CYCLE COST

Life cycle cost analysis helps you compare different technical solutions by contrasting the initial cost with operation, maintenance and replacement costs during the lifetime of the system. If we only compare the initial cost of an SPV system, we are ignoring all the recurring costs that need to be met in the future. The initial cost of an SPV system is relatively high, so it's important to take into account the operational savings the system can offer and quantify them economically.

The Directors of the community centre have realized that the increase in the price of fuel will mean higher and higher recurring costs if they use a diesel generator. The initial investment will be greater for an SPV system than a system using a diesel generator, but how do we compare the cost of the two options in the medium and long term, bearing in mind that the value of money and fuel change over time? Life Cycle Cost analysis provides a means of doing this, and includes all the costs an installation entails for the duration of its useful life (Figure 79).

Life Cycle Cost analysis allows you to work out the present value of a system including all the future expenditures. Life cycle cost is expressed as Net Present Value (NPV). Any analysis of this kind is dependent on a series of assumptions about what will happen in the future, and therefore includes a certain level of uncertainty. However, if used correctly, it's a useful tool for working out which option makes most economic sense.

Figure 78: Portable diesel generator

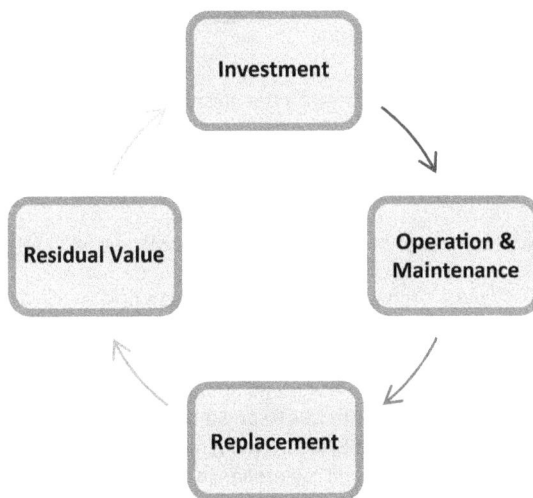

Figure 79: Stages of Life Cycle Cost analysis

A Life Cycle Cost analysis for the case study in question is presented below, comparing the SPV system with what the Directors were initially considering, a system with a diesel generator. To make the comparison between the two solutions and determine the life cycle cost, we must first define the following parameters:

➢ **The discount rate (opportunity cost, or cost of capital):** this reflects the present value of a future payment, and may also reflect the interest rate that a bank takes on a loan. The discount rate allows us to calculate the current value of a future payment

➢ **Energy inflation:** the general increase in energy prices in relation to the national currency during the period of analysis. In this case, the energy inflation we are interested in is the inflation in the price of diesel. Looking back over the figures for the last 10 years in Guatemala, we calculate an annual energy inflation rate of 6%.

➢ **Period of analysis:** the period of time covered by the analysis. In this case, we define it as 20 years, equivalent to the useful life of the SPV system.

➢ **Investment:** the initial cost of an installation. In this case, we have calculated that the total investment for the installation of a diesel generator is US$ 4750. For the SPV, it is US$ 9000.

➢ **Operation and maintenance:** the operating costs refer to the recurring expenses, in this case, costs derived from generating energy. For the SPV system, we calculate the operating cost as 0.00 US$/year, because we do not have to buy fuel in order to generate electricity. For the diesel generator, we need to work out the annual energy consumption in kWh (from the demand calculation in Spread Sheet 2 – Demand), the price of fuel in US$/litre (on the local market) and the fuel consumption of the generator in litres/kWh (using manufacturer's figures). The calculations are as follows:

$$\text{Anual demand (kWh)} = \frac{\text{Daily demand (Wh/day)}}{1000} = \frac{908 \text{ Wh x } 365}{1000} = \textbf{331 kWh/year}$$

The price of the electricity generated by the diesel plant is calculated as below:

$$\text{Price of Electricity (US\$/kWh)} = \frac{\text{Price of deisel (US\$/Litre)}}{\text{Generator consumption (Litres/kWh)}} = \frac{\text{US\$1}}{0.5 \text{ l/kWh}} = \textbf{0.5US\$/kWh}$$

The annual operating cost of the diesel generator is calculated as follows:

Operatio costs (US$/year) = Anual demand (kWh/year) x Price of electricity (US$/kWh)
= 331 kWh/year x 0.5 US$/kWh= 116 US$/ year

Maintenance cost refers to the on-going costs of maintaining an installation. For the SPV system, we estimate this to be **50 US$/year**. For the diesel generator, we estimate this to be **534 US$/year** (changing filters and oil, and engine servicing). The sum of the operating and maintenance costs is 166 + 534 = 700 USD$/year for the diesel plant, and 0 + 50 = **50 USD$/year** for the SPV system.

> ➢ **Replacement:** this refers to the cost of replacing parts of an installation, beyond regular maintenance. The diesel generator will need replacing once every 10 years (**twice** over the course of the 20 year analysis period), at a cost of **USD$ 3000** each time. For the SPV system, batteries will need replacing every 4 years (**5 times** over the analysis period) at a cost of **USD$ 1020**.

Table 21 shows the parameters that we will be using for the study, as defined above:

General parameters		
Discount rate	5 %	
Energy inflation	3 %	
Analysis period (years)	20	
Investment	**Diesel**	**SFA**
Initial cost (US$)	4750	9000
Operation and maintenance (O&M)	**Diesel**	**SFA**
Operating cost (US$/year)	166	0
Maintenance cost (US$/year)	534	50
Replacement	**Diesel**	**SFA**
Replacement diesel generator (USD$)	3000	-
Replacement SPV batteries (USD$)	-	1020
Number of replacements in 20 years	2	5

Table 21: Parameters for life cycle cost analysis

To calculate the life cycle cost, we have to add the cost of the initial investment to the net present value of the operation, maintenance and replacement costs for the duration of the system's useful life:

Life Cycle Cost = Investment + Operation (NPV) + Maintenance (NPV) + Replacement (NPV)

To calculate the present value of the future operating costs, we use the following formula:

$$\mathbf{NPV}_0 = A \times \frac{(1+e)}{(i-e)} \times \left\{ 1 - \left(\frac{1+e}{1+i} \right)^N \right\}$$

Where:

NPV$_0$ = Net Present Value (Operation)
A = Annual operating cost
e = Energy inflation
i = Dicount rate
N = Year

To calculate the present value of the future maintenance costs, we use the following formula:

$$NPV_m = A \times \left\{ \frac{(1+i)^N - 1}{i(1+i)^N} \right\}$$

Where:

NPV_m = Net Present Value (Maintenance)
A = Annual operating cost
i = Discount rate
N = year

To calculate the present value of the future replacement costs, we use the following formula:

$$NPV_r = \frac{A}{(1+i)^N}$$

Where:

NPV_r = Net Present Value (Replacement)
A = annual operating cost
i = Discount rate
N = year

Worked Example 16: Net present value of operation and maintenance costs (NPVo&m)

Diesel generator: using the parameters defined in Table 22, we calculate the NPVo for the diesel plant as follows:

$$NPV_0 = A \times 166 \times \frac{(1+0.06)}{(0.05-0.06)} \times \left\{ 1 - \left(\frac{1+0.06}{1+0.05} \right)^{20} \right\} = 166 \times 22.13 = \textbf{US\$ 3662}$$

SPV: we have determined that there are no operating expenses (fuel) for the SPV. Therefore, for the SPV:

$$NPV_0 = US\$ 0$$

Diesel generator: we calculate the NPV_m for the diesel plant as follows:

$$VAN_m = 534 \times \left\{ \frac{(1+0.05)^{20}-1}{0.05\,(1+0.05)^{20}} \right\} = 534 \times 12.46 = \textbf{US\$ 6655}$$

SPV:

$$VAN_m = 50 \times \left\{ \frac{(1+0.05)^{20}-1}{0.05\,(1+0.05)^{20}} \right\} = 50 \times 12.46 = \textbf{US\$ 623}$$

Diesel generator: NPVo&m is the sum of NPV_O and NPV_m = 3662 + 6655 = US\$ 10 317

SPV: $NPV_{O\&m}$ is the sum of NPV_O and NPV_m= **US\$ 623**

Worked Example 17: Net present value of replacement costs (NPV$_r$)

Diesel generator: the diesel plant has to be replaced at the end of years 10 and 20. We calculate the NPV$_r$ in two steps:

$$NPV_r = \frac{A}{(1+0.05)^{10}} = 3\,000 \times 0.61 = \text{US\$ 1 842}$$

$$NPV_r = \frac{A}{(1+0.05)^{20}} = 3\,000 \times 0.38 = \text{US\$ 1 131}$$

NPV$_r$ = 1 842 + 1 131 = US\$ 2 972

SPV: The SPV system batteries have to be replaced every four years, that is, at the end of years 4, 8, 12, 16 and 20:

$$NPV_{r1} = \frac{A}{(1+0.05)^{4}} = 1\,020 \times 0.82 = \text{US\$ 839}$$

$$NPV_{r2} = \frac{A}{(1+0.05)^{8}} = 1\,020 \times 0.68 = \text{US\$ 690}$$

$$NPV_{r3} = \frac{A}{(1+0.05)^{12}} = 1\,020 \times 0.56 = \text{US\$ 568}$$

$$NPV_{r3} = \frac{A}{(1+0.05)^{16}} = 1\,020 \times 0.46 = \text{US\$ 467}$$

$$NPV_{r4} = \frac{A}{(1+0.05)^{20}} = 1\,020 \times 0.38 = \text{US\$ 348}$$

$$NPV_r = 839 + 690 + 568 + 467 + 348$$

Worked Example 18: Life cycle cost as net present value (LCC$_{NPV}$)

LCC_{NPV} = Investment + $NPV_{o\&m}$ + NPV_r

Diesel generator:

LCC_{NPV} = 4 750 + 10 317 + 2 972 = **US\$ 18,039**

SPV:

LCC_{NPV} = 9 000 + 623 + 2 949 = **US\$ 12,572**

Figure 80 shows the results of the Life Cycle Cost study for each option. The investment cost of the SPV system is almost double that of the diesel generator. Nevertheless, the net present value of the operation and maintenance cost is much lower than the diesel option (US\$ 623 for the SPV and US\$10,317 for the diesel generator). Replacement costs are similar for both options. The results show that investing in an SPV system is preferable to the diesel generator option.

11.3 CALCULATING CO_2 EMISSIONS SAVINGS

Another important criterion to calculate is the savings in CO_2 emissions arising from an SPV system compared with the diesel option. Following on from the previous example, we can compare the CO_2 emissions from the operation of the diesel plant and the SPV system over the period of analysis (20 years). We have to establish the following criteria:

➢ **CO2 emissions factor:** the emissions factor determines how many kilos of CO_2 are emitted per kWh of energy generated by a system. The diesel generator emits on average **1 kgCO$_2$/kWh** of electrical energy generated. The SPV emits **0 kgCO$_2$/kWh**.

➢ **Annual energy demand:** from the demand calculation carried out in the sizing phase: **331 kWh**.

➢ **Analysis period:** 20 years.

Once we have defined the above parameters, we can calculate the CO_2 emissions per year, and the total for the analysis period. The results are shown in Table 22:

Life Cycle Cost

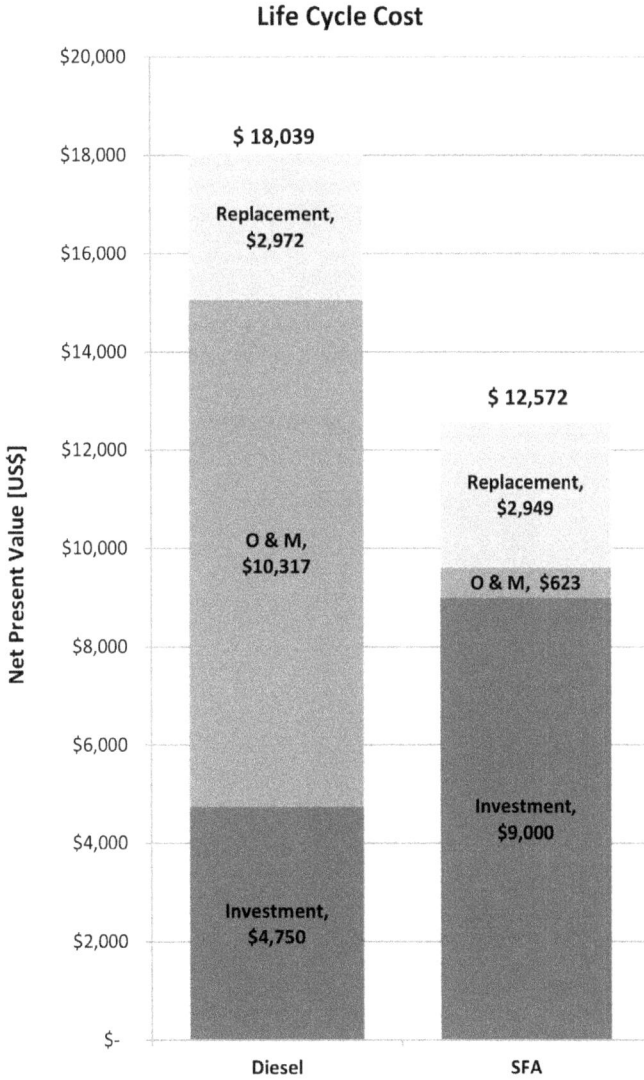

Figure 80: Results of life cycle cost analysis

	Diesel	SPV
CO_2 emissions factor diesel (kgCO_2/kWh)	1	-
CO_2 emissions factor SPV (kgCO_2/kWh)	-	0
Annual electricity demand (kWh/year)	331	331
CO_2 emissions (kg/year)	331	0
Analysis period (years)	20	20
CO2 emissions over useful life (kg)	6620	0

Table 22: Calculating CO_2 emissions

Figure 81 shows the results graphically.

The results indicate that the diesel plant will emit **6620 kgCO$_2$** over the life cycle, while the SPV will emit **0 kgCO$_2$**. The SPV, as well as having a lower life cycle costs, is a more environmentally friendly solution.

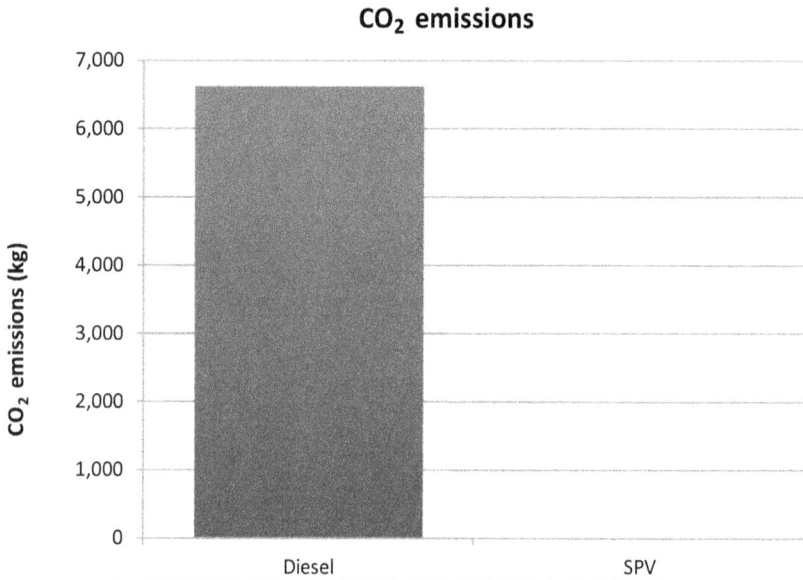

Figure 81: CO_2 emissions results

12 INSTALLATION AND COMMISSIONING

The most important points regarding the installation of each part of an SPV have been discussed in Chapters 3 – 9. The following includes general guidance for installing a low power SPV system. If you're installing a larger system, consult the technical resources in Appendix 16.1, find an expert in your area and follow the regulations and code in your country for installing SPV and electrical systems. Make sure the planning, sizing, design, and costing of the system is fully complete before preparing for installation.

Follow the guidlines below:

> **Preparation:** make a list of all the components, tools and materials needed for the installation, and check that you have everything before transporting it to site. SPV systems are often installed in remote places. If you are missing a screw you have a long journey before you can finish the installation: take spares! Make sure you have all the necessary technical information and documentation for the install.

> **Read the manuals** of all the parts you have bought and follow the manufacturers' recommendations regarding installation.

> **Work safely:** use common sense and don't work in damp conditions. Follow the regulations and recommended procedures in your country. Work calmly and avoid taking unnecessary risks. If you have no experience with electrical installations, get a qualified person to do the work.

> **Maintenance and enlargement:** an SPV system requires limited but regular maintenance in order to keep functioning properly. It's also possible that the system may be expanded in future. A high quality installation will make maintenance and enlargement easier!

➢ **Electronics are sensitive!** Many parts of an SPV system are sensitive to high and low temperatures and moisture. Carry out the installation in such a way that these temperature and moisture have the lowest possible impact on the useful life and functioning of the components.

➢ **Polarity:** with all DC circuits, ensure that you make the connections with the correct polarity so as to avoid damaging the equipment when the system is switched on. Double check all connections and polarities before commissioning.

Every installation is different. It depends on the specific system and where it is located. A general recommendation of the steps to be followed when installing a low power SPV is presented in Table 24 . The order in which they are carried out may vary.

Figure 82: Transporting the equipment for an SPV installation in a remote area

Before getting started with the installation, make a list of the steps that you are going to take and draw up a work plan for each day. This will help you to estimate how long the installation will take. Show the work plan to the installation team and get them to check it through. Reach a clear agreement on the work plan with the team before starting.

1 • After arriving on site, check off all the equipment, materials and tools on a checklist.

2 • Check where the modules, batteries, controller and inverter will be installed.

3 • Prepare the cables for the modules and connect them to them to the modules, taping off the other ends with electrical tape; prepare the mounting structure, and mount the modules.

4 • Prepare the cables for the charge controller and install the controller.

5 • Prepare the cables for the batteries (together with terminals) and install the batteries, taking the necessary safety precautions.

6 • Prepare the cables for the inverter and install the inverter.

7 • Prepare the cables for the loads and the ground connection, together with fuses and circuit protections.

8 • Complete all final connections and do a final visual inspection; test and commissions the system, in preparation for handover.

Table 24: General procedure for an SPV installation

12.1 Tools

The most important tools for an installation are listed in Appendix 16.1. Use high quality tools and treat them with respect. Make sure that they are looked after and stored safely when not in use.

An installation can be carried out much more quickly with a portable 12 V drill. Make sure the batteries are fully charged before you go. Using AC drills means that you have to take a portable electric generator, if there is no existing AC energy source. Depending on the budget available for tools, an ammeter is a very useful tool for an installation (Figure 83). It fulfils the same functions as a multimeter (measuring voltage in DC and AC) but allows you to measure current without disconnecting wires, as well as other functions, depending on the model.

For most installations a ladder will be indispensable: make sure that it is in good condition.

Figure 83: Ammeter

Figure 84: Tools for system maintenance

12.2 SAFETY

Throughout the course of this book we have included sections relating to safety issues when it comes to working with SPV systems. If you are an installer and you follow these recommendations you will be able to work safely with low power SPV systems. Don't take unnecessary risks! This is particularly important when installing SPV systems in remote places, far from the nearest medical centre. Below are some general recommendations for working with electricity and SPV systems:

GENERAL RECOMMENDATIONS:

➢ Have a first aid kit ready for treating minor injuries and locate the nearest medical centre to the installation site. All installers should have completed a first aid course.

➢ Before starting an installation, get the work team together, go over the safety regulations and agree on a procedure to follow in case of accidents.

➢ Everyone working on site should take off rings, bracelets, watches, necklaces and any other metal accessories before starting work.

➢ Use work boots with thick soles: they will protect your feet and insulate you in case you get an electric shock

➢ Never work with an SPV during a storm or in the rain.

➢ Use tools with electrical insulation.

➢ Always use insulation tape to insulate loose wires connected to modules and batteries while making connections.

➢ Do not work with live circuits: always take out the fuses or disconnect the load. Make sure that nobody can reconnect the circuits while you are doing the work.

> Most accidents happen when working at heights, installing modules or making wire connections. Use ladders that are in good condition and position them safely, and always have one person holding the ladder from below.

MODULES:

> These are the most expensive part of an SPV. Make sure they are well protected during transport.

> Putting up modules almost always involves working at heights with a piece of equipment that generates electricity: take the necessary precautions to prevent falls and risks of electrocution.

> Use a protective helmet if you are installing large modules, and wear a safety harness.

> Put boards on the roof so that you can move across roofs and ceilings more safely and so as not to damage the roof structure.

> To prevent the modules from generating energy during the installation, cover them with a thick rug or blanket.

BATTERIES:

> These are the most dangerous part of an SPV. Before starting to work with batteries, take off jewellery and metal accessories, wear protective clothing, gloves and goggles, take off your tool belt and make sure you have water and bicarbonate of soda nearby in case of an acid spill. Battery acid can cause blindness, destroy clothing and cause burns if it comes into contact with skin.

> When transporting batteries, make sure they are kept in an upright position with the terminals well insulated and nothing else placed on top.

> A metal key falling on the battery terminals can cause a short circuit, generating currents of thousands of amps. The key may melt and the batteries could explode. Only use insulated tools to make connections to a battery and leave tools on the floor when not in use.

> Batteries for SPV systems tend to be very heavy: to pick them up, use the straps or hold them from underneath, never by the terminals

> Never smoke near flooded liquid electrolyte batteries, and always install them in a well-ventilated spot away from the controller and inverter, as these can generate sparks and cause an explosion.

> The installation, commissioning and maintenance of a battery bank for larger systems SPV should only be done by qualified personnel.

12.3 INSTALLING MODULES

The modules are the parts of an SPV which generate energy and therefore it is very important that they are positioned correctly, at the right orientation and angle of inclination, to maximise the energy they generate throughout the year. You will have made a sketch of the site during the initial site visit, which will help you decide on an optimal location for the generator. Below is some general advice for installing modules:

> ➢ Maximise the solar irradiation captured: install the modules where they will receive the maximum amount of solar irradiation, making sure their orientation and angle of inclination are optimal for the latitude of the site (see Chapter 2).

> ➢ Shading, dust and dirt: install the modules in a place where there is no shading and where there is the least possible amount of dust (far away from stoves, chimneys and dust tracks or roads).

> ➢ Ventilation and cooling: make sure that air can circulate around the modules to minimise overheating.

> ➢ Voltage drop: reduce the distance of cable runs between components, to minimise conductor losses.

> ➢ Wind and protection from theft: make sure that the structure is capable of withstanding the strongest winds that could occur at the site. If there is a risk of theft, take preventative measures.

When you come to install the modules, remember that they will be there for 20 years: do it right first time!

If one part of a module is shaded (by a tree branch, for example) the power will be considerably affected. If this persists for a long time, it can damage the module.

12.4 MODULES: MOUNTS

Fixed module mounts keep them in a steady position throughout the year, at the optimal orientation and angle of inclination for your site. The structure needs to be strong, resistant to wind and moisture, and accessible so that the users can carry out maintenance and clean the modules.

Mounting modules directly on top of roofs can be problematic for SPV systems, especially in warm climates. If insufficient space is left between the modules and the roof, the modules will not get any ventilation, cell temperature will increase, and they'll produce less energy. This is particularly problematic with corrugated iron roofs: the metal gets hot during the day and reflects solar radiation, increasing the temperature of the cells and hindering ventilation Figure 85.

Figure 85: Avoid installing modules directly on top of galvanised metal sheeting!

Figure 86: Connecting wires for modules

Furthermore, it is generally quite difficult to alter the slope of a roof on a building, and even more difficult to rotate the entire building to get the right orientation. If there is a tree or neighbouring building that casts shade on the roof, a lot of energy will be lost over the course of the year and the modules could be damaged if only one or two cells are shaded. For low power SPV systems, it is best to install modules using a prefabricated metal pole mount. The advantage of this type of mount, is that it allows you to situate the modules where there is no shading, and at the optimal orientation and angle of inclination, making maintenance and cleaning easier.

Give the mount manufacturer the exact module dimensions and the location of the mounting holes. Make sure all metallic surfaces are painted with anti-corrosive paint, to protect them against the elements (if the metal is galvanized then the paint is not necessary). Figure 87 and Figure 88 shows an example of a pole mount.

You can see that the metal support allows the modules to be situated in the optimal location in a wooded area where the trees cast a lot of shade. The orientation and angle of inclination of the roof are unsuitable for capturing solar irradiation at this site. A cheaper alternative is to build a module support out of aluminium and fix it onto a solid wooden post with clamps or braces. Bear in mind that the post will have to be changed after 10-15 years, depending on the kind of timber.

12.5 MODULES: CONNECTIONS IN PARALLEL AND SERIES

Module connections are made depending on the number of modules, their voltage and the system voltage. Figure 89 shows an example of modules connected in series and in parallel.

The circuit on the left shows two 12 V modules connected in parallel for a 12 V system, providing a system voltage of 12 V. In the circuit on the right, two 12 V modules are connected in series, for a 24 V system.

The junction box where wires are connected, is found on the back of the module. Series connections are usually made inside the junction boxes of the modules themselves. Parallel connections of several rows connected in series are usually made in a separate junction box, with blocking diodes and other protective elements specified in the design.

During the installation, make sure you isolate the cables that are connected to the modules. If the array is small, cover it with a blanket so that the modules do not produce energy until the final connections have been made.

Figure 87: Metal pole mount for modules

Figure 88: Aluminium frame and solid wood pole mount

12 V PARALELL CONNECTION   **24 V SERIES CONNECTION**

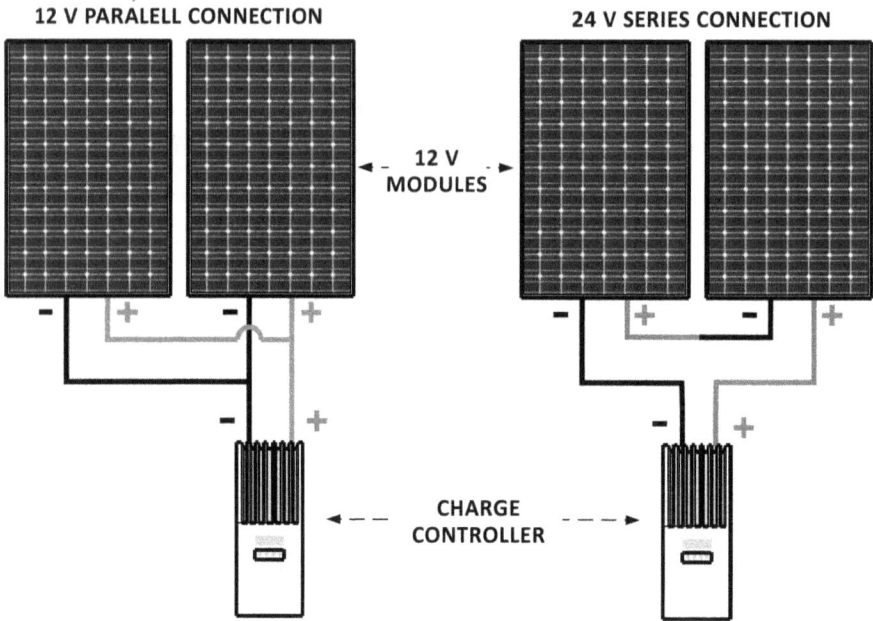

Figure 89: Connecting modules in series and in parallel

12.6 INSTALLING BATTERIES

The batteries are very costly over the life cycle of the system, as they have to be replaced three or four times over the system´s lifetime (depending on the type of battery, the usage regime and the quality of maintenance). Below are some general recommendations for installing batteries:

➢ **Protection against the elements:** make a space for the batteries that is secure, clean, and protected from outside weather conditions, but which facilitates maintenance. Never install batteries where they will be exposed to direct sunlight. The optimal operating temperature for a lead-acid battery is 25º C (77º F).

➢ **Ventilation and spark protection:** flooded liquid electrolyte batteries should be installed in a ventilated space, away from the other components (controller and inverter) and where there is no risk of sparks or flames.

➢ **Battery box:** in certain situations, it will be necessary to build a small outhouse exclusively to house the battery bank.

➢ **Stand:** batteries should be installed on a raised platform, protected against corrosion and battery acid. This is important to minimise self-discharge of the batteries.

➢ **Inclination:** install the batteries on a perfectly horizontal surface.

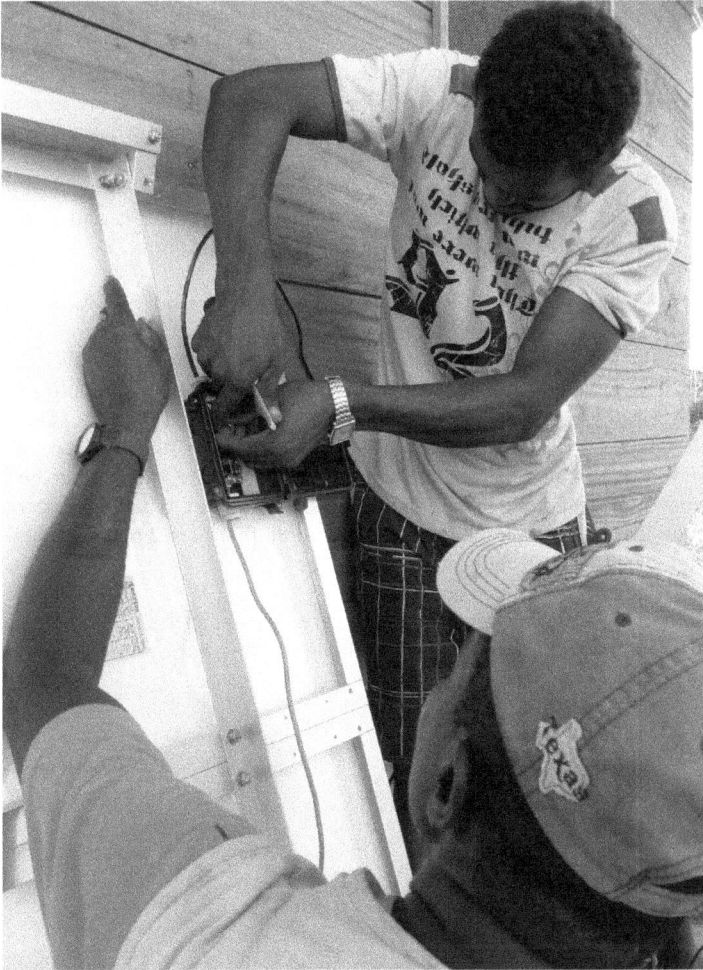

Figure 90: Making the connections on a photovoltaic module

Figure 91: Clamp-Fix Battery Terminal

Figure 92: Screw-Fix Battery Terminal

➢ **Ventilation:** leave a space of at least 3cm between batteries to allow air circulation.

➢ **Terminals and connection:** use high quality terminals to connect the wires to the batteries, clamp or screw on (Figure 91 and Figure 92). Make sure the polarity is correct and that all terminals are well secured. Apply petroleum jelly to the terminals after making the connections to prevent sulphation.

➢ **Voltage drop:** keep the distances between the batteries and inverter and/or controller to a minimum, to avoid excessive voltage drops.

➢ **Protections:** fuses for the batteries must be installed on the positive wire, as close as possible to the positive battery terminal.

The most suitable wires to connect the batteries to the controller and/or inverter are copper wires, for welding (see Chapter 8). Flooded liquid electrolyte batteries are often sold dry: the acid comes separately in sealed containers, to make transport easier. When the batteries and containers of acid have reached the installation site, they need to be filled using a clean funnel or an acid pump (available from a solar supplier). When the batteries are full, leave them for 10 minutes, and fill them up again until there are 2 or 3 centimetres of free space left between the electrolyte and the ventilation cap.

The battery connections depend on the battery voltage and the system voltage required. Figure 93 , Figure 94, and Figure 95 show examples of different configurations, with series, parallel and mixed connections.

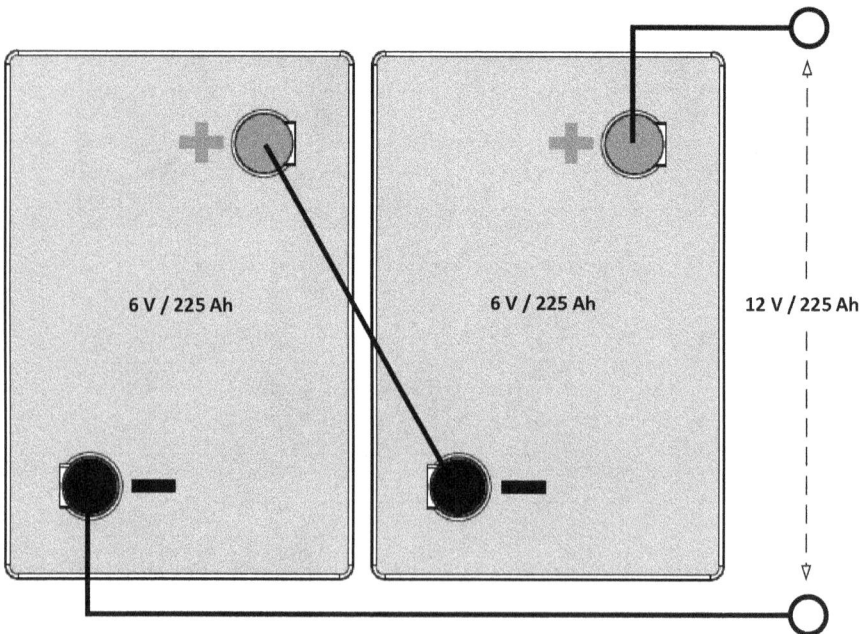

Figure 93: Two 6V batteries connected in series for a 12 V system

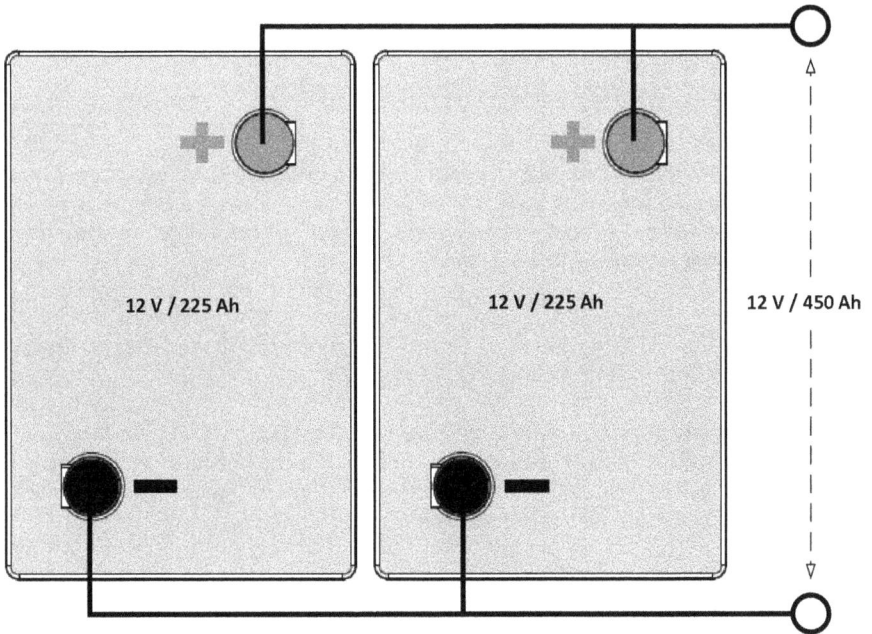

Figure 94: Two 12 V batteries connected in parallel for a 12 V system

12 V / 225 Ah

12 V / 225 Ah

12 V / 450 Ah

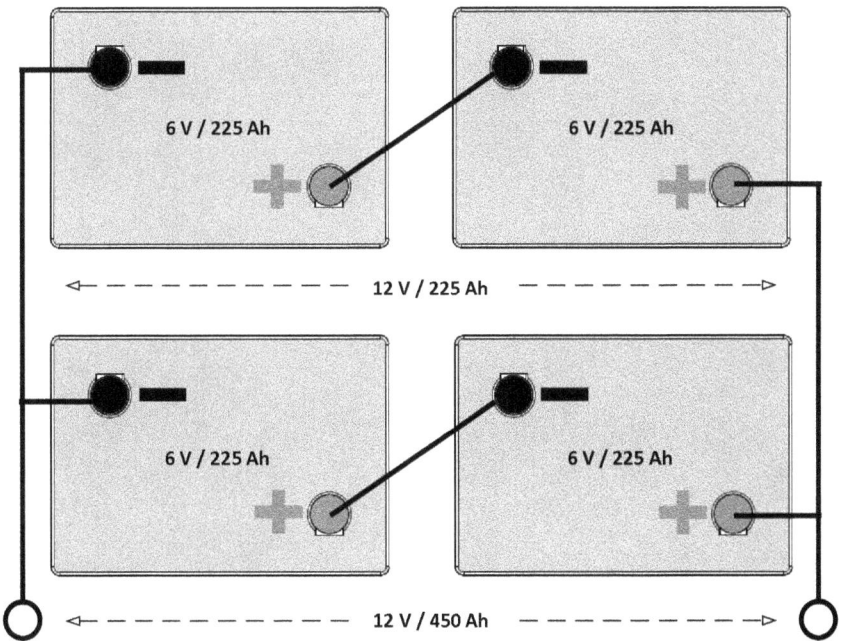

6 V / 225 Ah

6 V / 225 Ah

12 V / 225 Ah

6 V / 225 Ah

6 V / 225 Ah

12 V / 450 Ah

Figure 95: Mixed connection of four 6 V batteries for a 12 V system

Before carrying out the installation, make a schematic of the battery bank connections, making clear the layout of batteries and cables. Installing a battery bank for larger systems is delicate work with greater risks: find a qualified person for the job.

Do not mix new batteries with old in a battery bank! All the batteries should be the same age, make and model. If you replace just one battery in a bank of old batteries, the new one will only work at the level of the old batteries and you will be wasting money.

12.7 INSTALLING THE CONTROLLER

The most important factor when it comes to installing the controller is the length of the wire run to the batteries. Almost all controllers come with anchorage holes which allow them to be fixed to a wall or similar surface with screws. Controllers usually have six connection terminals for all the circuits, and their polarity is indicated.

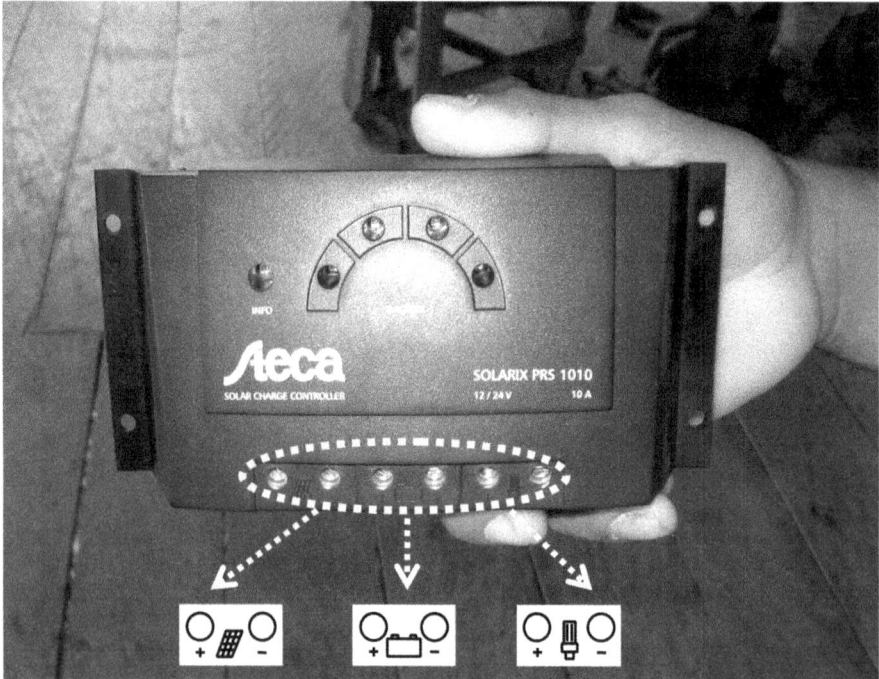

Figure 96: Connection terminals on a controller

Table 23 shows the connection terminals that are typically found on a controller:

Terminales de conexión de un regulador	
	This indicates the terminal block for module connection, with the positive (+) and negative (-) terminals.
	This indicates the terminal block for battery connection, with the positive (+) and negative (-) terminals.
	This indicates the terminal block for DC load connection, with the positive (+) and negative (-) terminals).

Table 23: Terminales de conexión de un regulador

The manufacturer's manual will provide the necessary information for installing and connecting the controller.

Never connect a module or appliance directly to a battery without going through the controller, as this can damage the battery. For an SPV system using AC, the inverter is never connected to the charger: it connects directly to the batteries. Do not equalise a sealed battery! Equalisation is only for flooded liquid electrolyte batteries.

12.8 INSTALLING THE INVERTER

If your system has AC loads, you will need to install an inverter. The most important factor in installing the inverter (as with the controller) is the length of the wire run to the batteries. Broadly speaking, inverters are designed to be installed indoors. Almost all SPV inverters have an electric fan, so it is very important that air is able to circulate freely around the inverter, to cool it down when running high loads.

An inverter is usually much bigger and heavier than a controller. However in most cases they have anchorage holes which allow them to be fixed to a wall or similar surface with screws. The manufacturer's instructions will show which fixing positions are allowed (vertical, horizontal etc.).

The connection points on the inverter will be shown in the manufacturer's manual. These generally consist of two connection terminals for incoming DC current (with their respective polarities), and two connection terminals for AC output (phase, neutral), as well as the earth terminals. Figure 98 shows an example.

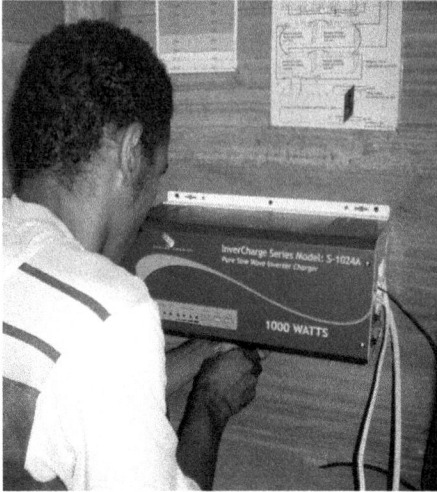

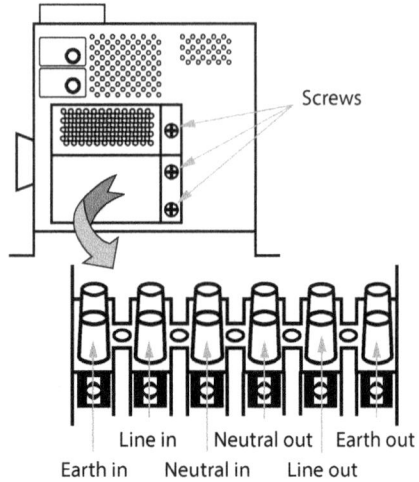

Screws

Line in   Neutral out   Earth out
Earth in   Neutral in   Line out

Figure 97: Mounting an inverter

Figure 98: Example of connection terminals on an inverter [Source: Xantrex]

The connection points on the inverter will be shown in the manufacturer's manual. These generally consist of two connection terminals for incoming DC current (with their respective polarities), and two connection terminals for AC output (phase, neutral), as well as the earth terminals. Figure 99 shows an example.

Due to the fact that an inverter works with higher voltages than a controller (outside the range that is considered safe for people), the connection terminals tend to be housed inside a special protective cover.

12.9 ELECTRICAL SCHEMATIC

For the benefit of the installation process and anyone who carries out maintenance on the system in future, draw up an electrical schematic which clearly shows all the parts of the system and wiring, cable cross-sectional areas, junction boxes, circuit breakers and other protective elements. An example is shown in Figure 100.

Leave copies of the schematic and other documentation securely fixed somewhere in plain view, to make it easier for the end users to manage and maintain the system.

12.10 INSPECTION, TESTING AND FINAL CONNECTIONS

Before making the final connections, inspect and test all parts of the system:

DC terminal caps

Figure 99: Protective covers for the DC terminals on an inverter [Source: Outback]

CIRCUIT DIAGRAM

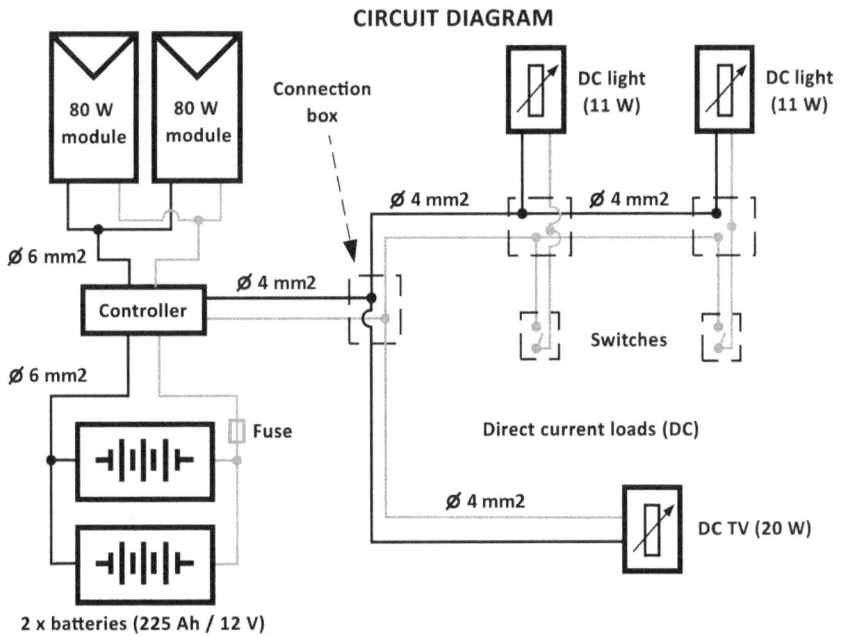

80 W module

80 W module

Connection box

DC light (11 W)

DC light (11 W)

Ø 4 mm2

Ø 4 mm2

Ø 6 mm2

Ø 4 mm2

Controller

Switches

Ø 6 mm2

Fuse

Direct current loads (DC)

Ø 4 mm2

DC TV (20 W)

2 x batteries (225 Ah / 12 V)

Figure 100: Example of an electrical schematic for a low power SPV system

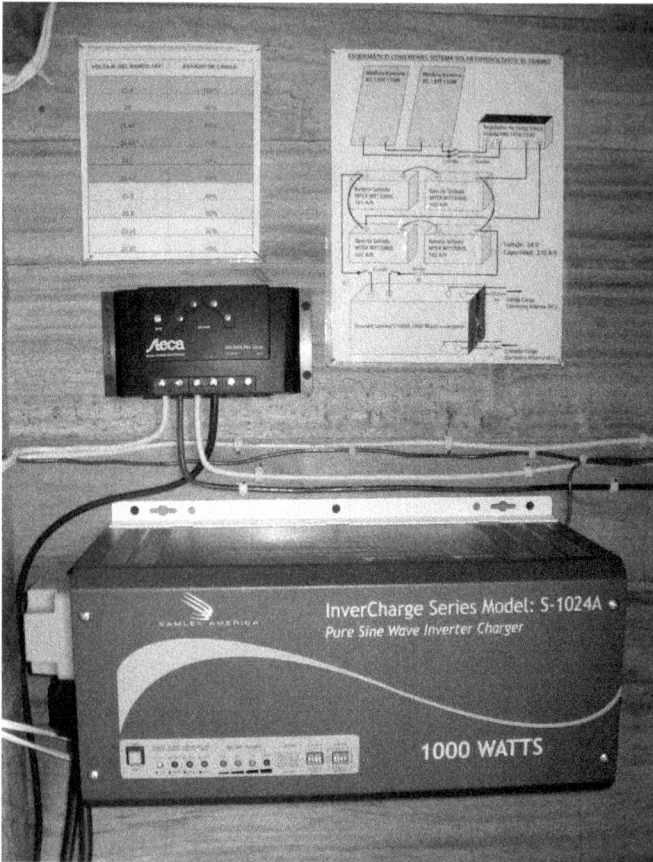

Figure 101: Installation with technical documentation on display

➤ Visually inspect the condition of the modules, support structure and wires.

➤ Visually inspect the condition of the batteries and electrolyte levels.

➤ Visually inspect the controller, inverter and appliances.

➤ Check that the modules are generating electricity, by measuring the open circuit voltage and current of the photovoltaic array using a multimeter or ammeter.

➤ Check the state of charge of the batteries.

➤ Make sure that all terminal connections and connections between circuits have been made properly, that terminal screws are well secured and that there is no risk of a wire coming loose.

➤ Check that all the wires in the DC circuits have the correct polarities.

> ➤ Check that all the protective elements (fuses, earth rod etc.) are correctly installed and connected. Test the continuity of the wires in the earth circuit.

> ➤ Test continuity in the distribution circuits with a multimeter and check that the switches are working properly.

Once the system has been inspected and tested, you can proceed with the final connections. When making the final connections, all circuit protections and switches should be set to the open position (disconnected). The recommended order in which to make the final connections is shown in Figure 102 .

| 1. Battery to charger | 2. Battery to inverter | 3. Charger and inverter to load | 4. Charger to modules |

Figure 102: Recommended sequence for final connections

Figure 103: Successful test of the lighting system running for an SPV

Figure 104: Checking that a hi-fi system works with a newly installed SPV system

12.11 COMMISSIONING

Commissioning an SPV is the last step after making the final connections. Commissioning systems with a capacity ≥ 500 Wp is a job for a qualified electrician with experience in photovoltaic systems. The process consists of making final checks and tests on the system to make sure that it is functioning correctly, so that you can formally hand the system over to the client. You need to measure voltage drop at the furthest points of all the circuits while the system is in full sunlight and the loads are connected. Check that all appliances are working properly. If an inverter is included in the system, check that appliances like computers and sound equipment work properly without making buzzing sounds.

The final handover of the system should be made when the batteries reach a 100% state of charge.

12.12 SYSTEM HANDOVER AND DOCUMENTATION

When the commissioning has been completed, proceed with the formal handover of the system. End-users or operators should be shown all the parts of the installation, with an explanation of how the system works, and the maintenance requirements.

Figure 105: Signing of handover document

Draw up and sign a document which states that the system has been handed over, including information about the warranties on the different parts of the system, and the installer's responsibilities in case of problems in the future.

The final handover document should be accompanied by all the relevant technical documents, including manuals, data sheets, preventative maintenance record sheets, plans, diagrams and schematics. When put together a project, be meticulous and organised with the project documentation from beginning to end, so that anyone who works on the system in future can quickly locate the information they require.

The final handover of an SPV system is the culmination of a lot of hard work. In remote areas, where there is little access to electrical energy, the handover is an important moment and a good reason to celebrate.

Figure 106: Celebrating with a film after the final handover of an SPV

13 OPERATION AND MAINTENANCE

The basic operation and maintenance required for an SPV system that has been specified and installed well is relatively simple. The most effective maintenance is always preventative, carried out regularly, according to a standard schedule, and recorded for control at a later date. This way, the system is kept in an optimal condition and operation and maintenance costs are reduced.

Generally speaking, SPV systems fail for two reasons: bad management by end-users/operators, or a fault due to problems with the design or installation. In both cases, the impact is usually seen first in the batteries. If end-users' energy management is poor, and they do not understand the capacity of the system and its limitations, they will be continuously discharging the batteries beyond the depth of discharge specified for the system, causing the batteries to fail prematurely. If a piece of equipment fails for some reason, other than those described above, check the warranty and contact the supplier for repair or replacement.

When you make the final handover of an SPV system in a remote area, hold a training session with the local operators or end-users. Go over all of the preventative maintenance tasks and their schedule (how many times per week, month, year etc.), using a maintenance record sheet (Figure 107), The operator/user can fill in the document, recording the maintenance tasks and helping identify problems before they get out of hand.

Some controllers are capable of storing data of the amount of energy received from the modules and consumed by the appliances (in DC) - a useful function for monitoring system performance and checking the efficiency of the design. For low power SPV systems, the cost of monitoring is usually prohibitive. However, for larger set ups, a monitoring system (with long-term data logging) is the best way of analysing system performance, year after year. Whenever possible, plan and budget for monitoring and evaluation of the system following installation. This provides useful feedback on whether the installation meets the design specifications and will improve future projects.

Location:	
Month/year:	
Operator:	

PHOTOVOLTAIC MODULE	Week 1	Week 2	Week 3	Week 4
Cleaning of panels with cloth, water, soap				
Inspection of connections, cables, terminals				
Inspection of module mount				
Observations				

BATTERIES	Week 1	Week 2	Week 3	Week 4
Cleaning of surface of batteries				
Inspection of connections, cables, terminals				
State of charge	Tension (V)	Tension (V)	Tension (V)	Tension (V)
(min. ½ hour at rest)				
Observations				

INVERSOR	Week 1	Week 2	Week 3	Week 4
Inspection of connections, cables, terminals, fan				
Observations				

REGULADOR	Week 1	Week 2	Week 3	Week 4
Inspection of connections, cables, terminals				
Observations				

Figure 107: Example of a maintenance record sheet

13.1 ENERGY MANAGEMENT IN AN SPV

The best kind of electrical energy is the nega-watt: the kind that's not consumed! Managing energy in an optimal way means minimising demand, reducing phantom loads, turning off appliances and lighting when they're not required, and reducing demand during cloudy weather conditions (if necessary).

It is important to eliminate all the phantom loads in an SPV. Phantom loads are drawn by inverters, charging devices or power supplies, for mobile phones, laptop computers, printers, etc. They consume a minimal amount of energy, despite the appliance being switched off. If there are several phantom loads in a system, connected day and night, the energy lost over a month becomes significant. Always disconnect charging devices or power supplies when they are not in use, or connect them to a power strip with its own switch, that cuts off power to several appliances at once (Figure 108). Timers can also be installed to make sure lighting and appliances only work at specific times.

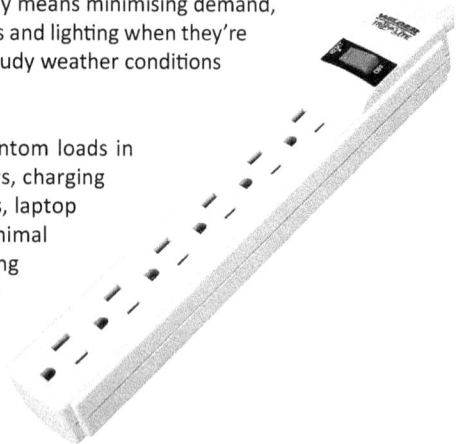

Figure 108: Multi plug extension lead

All inverters consume energy in stand-by mode when there is no load connected. Turn off the inverter when it's not in use! Some inverter models automatically go into rest mode when no loads are detected, which reduces stand-by energy consumption.

It's important that end-users or operators continue using efficient appliances and light bulbs with an SPV system. When a light bulb breaks, it is important to replace it with one that is just as, or more efficient. Repainting walls and ceilings with white paint, boosts brightness levels: this is a much more efficient solution than installing more light bulbs or installing bulbs with a higher light output.

If the system is to be enlarged, redo the demand calculations, specify the modules which will be added, size the new battery bank, and then carry out the enlargement, always specifying efficient equipment. Remember that you should never buy new batteries and mix them with old batteries: the new ones will not work at full capacity and the money won´t be well spent.

13.2 BATTERIES

Batteries are the cornerstone of the system, and require careful management and maintenance. The higher their average state of charge, the longer they will last, and the lower the maintenance and replacement costs will be for the system as a whole. They should reach a 100% state of charge at least once a week. Before working with batteries, take the safety precautions listed in Chapter 12. If you're working with flooded liquid electrolyte batteries, you need to apply an equalization charge every month, and refill them with distilled water, checking the electrolyte level weekly.

Check the batteries' state of charge every week and write it down on the maintenance record sheet. Unless you have a controller that includes an electronic compensation function or a dedicated monitor for measuring the state of charge, the batteries need to be left without charging or discharging for at least half an hour to get an accurate reading (for example, first thing in the morning).

The recommended procedure for cleaning batteries is as follows (monthly, or as required):

➢ Disconnect the load from the modules.

➢ Take the fuse out and remove the terminals from the contact points on each battery.

➢ Carefully clean the outside surfaces of the batteries, paying particular attention to the top, which is where collected dirt causes self-discharge.

➢ Clean the contact points with sandpaper (if they are made of lead) until they shine. If sulphation has occurred, apply a solution of water and caustic soda after cleaning. If the contact point has become extremely corroded it will need replacing.

➢ Fix the terminals on again and tighten the bolts. Lastly, apply grease or petroleum jelly to prevent sulphation.

13.3 MODULES

Modules require minimum maintenance, but the most important task is to clean the glass covering the cells. Make a visual inspection of the area surrounding the array, and make sure that there are no tree branches casting shade on the modules. Modules should be cleaned more frequently (every two weeks) during dry months, using a clean soft cloth and soapy water.

Every 6 months you need to check that the mount is in good condition, and apply anti-corrosive paint if necessary. Check the junction boxes on the back of all the modules to be sure insects haven't made a home there. Make sure that the wires are all in good condition and firmly secured- if they weren't properly installed, they can be loosened from the terminals following heavy winds. Measure the modules' voltage and current every year under full midday sun, and record the results on the maintenance record sheet. Compare the results year on year so that you can establish whether the modules are achieving the performance specified in the warranty.

13.4 CONTROLLER, INVERTER, AND ELECTRICAL INSTALLATION

The controller and inverter require minimal maintenance: check the wires and terminals (by gently tugging on cables), clean the surfaces, and make a visual inspection of the LED indicators and inverter fan, to make sure that nothing is stopping the air from circulating.

Ever year the electrical installation requires a visual inspection of the wires and wire insulation, terminals, junction boxes and circuit breakers. If a fuse burns out or a trip switch is activated while the system is operational, it will be immediately noticeable and should be fixed. Check the earth rod and the wires in the earth circuit. Clean all the light bulbs regularly with a damp cloth (always switch them off first). If the tube of a fluorescent light bulb begins to turn black, it will be coming to the end of its useful life and will need replacing.

14 GLOSSARY OF TERMS

AMP	*Unit used to measure the intensity of an electrical current.*
AMP-HOUR	*Unit used to specify the charge capacity of a battery.*
ANGLE OF INCLINATION	*The angle formed between a photovoltaic panel and a completely flat or level surface.*
ARRAY	*A collection of photovoltaic modules that generate electric energy.*
AUTONOMOUS, STAND-ALONE OR OFF-GRID SYSTEM	*Photovoltaic system that works with no other source of energy and is not connected to the grid, requiring batteries or other means of storage.*
BATTERY	*Component with the capacity to store electrical energy, transforming it into chemical energy, and vice versa.*
BATTERY BANK	*Several batteries connected together in series or in parallel, with the capacity to store electrical energy, transforming it into chemical energy and vice versa.*
CHARGE CONTROLLER	*A component that controls and regulates the charging process of the batteries in an SPV system.*
CHARGE REGULATOR	*A component that controls and regulates the charging process of the batteries in an SPV system.*
DISCOUNT RATE	*The discount rate, opportunity cost or capital of cost; a financial parameter used to the net present value of a future payment.*
ELECTROLYTE	*A solution of diluted sulphuric acid, within which the various processes take place that allow the battery to charge and discharge cyclically.*
ENERGY	*The capacity to do work. Energy = power x time.*

ENERGY EFFICIENCY	The ratio of input energy to output energy, usually measured as a percentage (%).
GRID CONNECTED PHOTOVOLTAIC SYSTEM	Photovoltaic system that feeds AC electrical energy into the electricity grid. This type of system consists of modules and an inverter for grid connection, but no batteries.
HYBRID SYSTEM	Photovoltaic system that includes other sources of electricity generation, such as wind turbines or generator plant .
INFLATION	The general increase in the price of goods and services in relation to a national currency over a given period of time.
INVERTER	Component that transforms direct current supplied by batteries or modules into alternating current for use in different electrical appliances or applications.
KILOWATT (kW)	Unit of power equivalent to 1000 Watts (W).
MAXIMUM POWER POINT OF A MODULE (MPP)	The power supplied by a photovoltaic module when the product of voltage and intensity is at its point of maximum power .
NET PRESENT VALUE	A unit that allows the present value of an investment to be calculated on the basis of given number of future cash flows.
NOMINAL VOLTAGE	Specific potential difference for which a piece of equipment or installation is designed. The term "nominal" refers to the fact that the voltage can vary under different operating circumstances.
NORMAL OPERATING CELL TEMPERATURE, (NOCT)	The temperature at which the cells of a photovoltaic module function under standard operating conditions (SOC)
OHM	Unit of electrical resistance.
OPEN CIRCUIT VOLTAGE	The potential difference measured between two ends of an electrical circuit when it is open and no load is connected.
PHOTOVOLTAIC (PV)	The process by which electric energy is generated from solar irradiation.
PHOTOVOLTAIC CELL	Basic unit of a photovoltaic module, where the transformation of sunlight to electrical energy occurs .
PHOTOVOLTAIC EFFECT	The conversion of light energy to electrical energy.
PHOTOVOLTAIC MODULE	Set of several interconnected photovoltaic cells, encapsulated and protected by glass and an aluminium frame.
POWER	The amount of energy delivered.
RESISTANCE	The opposition met by an electrical current as it passes through a material.
SILICON	Chemical element of which the cells of a solar panel are basically constituted. It is dark grey and almost metallic in appearance and has excellent semiconductor properties.
SOLAR AZIMUTH	The angle of orientation from the North-South line on a horizontal plane, measured in degrees. South corresponds to 0º, East -90º, West 90º, and North 180º

SOLAR IRRADIANCE	Measure of the amount of solar energy that falls on a given surface over a given period of time, measured in W/m², kW/m²
SOLAR RADIATION	The amount of solar energy that hits a given surface over a given period of time
STANDARD OPERATING CONDITIONS (SOC)	The laboratory test conditions used for testing a photovoltaic module:
STANDARD TEST CONDITIONS (STC)	The laboratory test conditions used when testing a photovoltaic module under the following conditions.
VOLT (V)	The potential difference at the end of a wire when a current of one amp uses one watt of power.
WATT (W)	Unit of electric power, equivalent to one joule per second.
WATT PEAK (Wp)	Unit of power which refers to the product of the voltage and intensity (peak power) of a photovoltaic module under standard test conditions (STC).

15 BIBLIOGRAPHY

Antony F., Dürschner C., Remmers K-H (2006), Fotovoltaica para profesionales, PROGENSA, Spain.

Asociación de la Industria Fotovoltaica (2005), Sistemas de Energía Fotovoltaica, PROGENSA, Spain.

Goss B. (2010), Choosing Solar Electricity, Centre for Alternative Technology, United Kingdom.

Hankins M. (2010), Stand-alone solar electric systems, Earthscan, United Kingdom.

Ingeniería Sin Fronteras (1999), Energía Solar Fotovoltaica y Cooperación al Desarrollo, IEPALA Editorial, Ingeniería Sin Fronteras, Spain.

Kerridge D., Linsley Hood D., Allen P., Todd B. (2008), Off the grid, Centre for Alternative Technology, United Kingdom.

McMullan R. (2007), Environmental Science in Building, Palgrave Macmillan, United Kingdom.

Northern Arizona Wind and Sun (2011), Deep Cycle Battery FAQ, Northern Arizona Wind and Sun, USA.

Pareja Aparicio M. (2010), Energía Solar Fotovoltaica, Cálculo de un instalación aislada, Marcombo, Spain.

16 APPENDICES

16.1 RESOURCES FOR SPV SYSTEMS

NASA Surface Meteorology and Solar Energy

Database of meteorological and solar radiation data for anywhere in the world.

http://eosweb.larc.nasa.gov/sse/RETScreen

Photovoltaic Geographical Information System (PVGIS)

Database of solar radiation for Europe and Africa and yield calculations for photovoltaic generators.

http://re.jrc.ec.europa.eu/pvgis/index.htm

University of Oregon Solar Radiation Monitoring Laboratory

Generate a sun path diagram for anywhere in the world.

http://solardat.uoregon.edu/SunChartProgram.html

Simulation and calculation programs for SPV systems

- ➢ RETScreen (free software): www.retscreen.net

- ➢ PVSYT: www.pvsyst.com

- ➢ PVSol: www.valentina.de

- ➢ HOMER: www.homerenergy.com

16.2 SPREAD SHEETS

SPREAD SHEET 1 – SOLAR RESOURCE

SPREAD SHEET	1 SOLAR RESOURCE
Prepared by	
Prepared for	
Date	
Location	
Coordenates	

Step	Solar irradiation on a horizontal surface (kWh/m2/día) - Peak Sun Hours												
	January	February	March	April	May	June	July	August	September	October	November	December	*Average*
1													

2	Critical month	

SPREAD SHEET 2 – ENERGY DEMAND

SPREAD SHEET		2 ENERGY DEMAND					
Step	Load	Quantity x	Power (W) x	Hours of use/day (h) =	DC demand (Wh/day)	o	AC demand (Wh/day)
1	Loads in Direct Current (DC)	x	x	=			
		x	x	=			
		x	x	=			
		x	x	=			
	Loads in Alternating Current (AC)	x	x			=	
		x	x			=	
		x	x			=	
		x	x			=	
2	Daily demand in DC (Wh)						
	Daily demand in AC (Wh)						
3	Percentage of losses in DC (%)						
	Percentage of losses in AC (%)						
4	Losses in DC (Wh) [daily demand x losses]						
	Losses in AC (Wh) [daily demand x losses]						
5	Corrected demand in DC (Wh/day) [daily demand x losses]						
	Corrected demand in AC (Wh/day) [daily demand x losses]						
6	Total daily demand (Wh/day) [corrected DC demand + corrected AC demand]						
7	System voltage (V)						
8	Required daily charge(Ah) [total daily demand ÷ system voltage]						

SPREAD SHEET 3 - MODULES

SPREAD SHEET		3 MODULES			
Step	**Daily charge (Ah)**	÷	**PSH critical month**	=	**System charging current (A)**
1		÷		=	

2	Module		
	Nominal voltage (V)		
	Maximum power (W)		
	Maximum current Isc (A)		
	Current SOC (A)		÷ ←

3	**Number of modules**	

SPREAD SHEET 4 – BATTERIES

SPREAD SHEET		4 BATTERIES				
Step	**Daily charge (Ah)**	x	**Days of autonomy**	÷	**Depth of discharge***	=	**Battery bank capacity (Ah)**
1		x		÷		=	

2	Battery	
	Nominal voltage (V)	
	Capacity @ C20 (A)	

3	Batteries in series		
	Batteries in parallel		
	Number of batteries		Equal: yes or no?
	Total capacity (Ah)		

4	**Useful life @ selected depth of discharge* (cycles)**	
	Estimated useful battery life (life)	

SPREAD SHEET 5 – CONTROLLER AND INVERTER

SPREAD SHEET		5 CONTROLLER & INVERTER		

CONTROLLER

Step	Max. current modules (A)	x	Safety margin (%)	=	Corrected maximum current modules (A)
1		x	25%	=	

Which larger?

Step	Max. DC load current (A)	x	Safety margin (%)	=	Corrected maximum current DC load (A)
2		x	25%	=	

3	Maximum current charge controller (A)	

INVERTER

Step	Max. power AC load (W)	x	Safety margin (%)	=	Max. corrected power AC load (W)
4		x	25%	=	

5	Minimum nominal power inverter (W)	

SPREAD SHEET 6 – WIRES AND FUSES

SPREAD SHEET		6 WIRING AND FUSES		

WIRING

Step	Circuit	Cable run length (m)	Max. current (A)	Wire resistance factor K (Ω/m)	Total resistance (Ω)	Voltage drop (V)	Voltage drop (%)
1							

FUSES

Step	Circuit	Max. power (W)	Max. current (A)	Nominal current fuse rating (A)
2				

16.3 ENERGY CONSUMPTION OF COMMON APPLIANCES

Appliance	Consumption (W/h)	Notes
Toaster	1000	Not suitable for SPVs
Blender / liquidiser	300	Not suitable for SPVs
Microwave	600 - 1500	Suitable for more powerful SPVs
Iron	1000	Not suitable for SPVs
Hairdryer	1000	Not suitable for SPVs
Electric fan	10 - 250	Suitable for more powerful SPVs
Washing machine	900	Not suitable for SPVs, unless they are very efficient
Tumble dryer	4000	Not suitable for SPVs
Laptop computer	20 - 50	Suitable for SPVs
Desktop computer	100 - 220	Not suitable for SPVs
Ink printer	50	Suitable for SPVs
Laser printer	600	Not suitable for SPVs
Electric drill	500	Not suitable for SPVs
Circular saw	1200	Not suitable for SPVs
DC fridge/freezer	100	Suitable for SPVs
AC fridge/freezer	500	Not suitable for SPVs, unless they are very efficient
Hi-fi system	10 - 300	Suitable for SPVs
Video player	10 - 30	Suitable for SPVs
DVS player	50	Suitable for SPVs
Sewing machine	100	Suitable for more powerful SPVs
Public frequency radio	5	Suitable for SPVs
DC television	20	Suitable for SPVs
AC television	70 - 150	Suitable for more powerful SPVs

16.4 Basic tools for installing an SPV system

Herramienta	Función
Compass	Working out the orientation of the generator
Tape measure	Measuring distances
Knife	Cutting various objects
Hygrometer	Measuring the state of charge of open batteries
Screwdrivers	Tightening screws and terminals
Inclinometer	Working out the angle of inclination of the modules
Lamp	Installing cables and equipment in dark places
File	Filing
Adjustable spanner	Tightening nuts and bolts on structures and battery terminals
Mallet / mace	Enterrar varillas de tierra
Hammer	Various tasks
Digital multimeter	Checking connections, measuring voltage
Spirit level	Various installations
Spade	Digging trenches and holes
Cable stripper	Preparing wires
Compression pliers	Preparing terminals
Hacksaw	Cutting metallic objects
Handsaw	Cutting wood
Pliers	Cutting wires, tightening nuts and bolts

16.5 Solar irradiation data for Latin America

Daliy solar irradiation on horizontal surface [kWh/(m2.d) or Peak Sunlight Hours]

City	Jan.	Feb.	Mar.	Apr.	May.	Jun.	Jul.	Aug.	Sep.	Oct.	Nov.	Dec.	Annual Ave.
MEXICO													
Aguascalientes	4.73	5.72	6.85	7.20	7.18	6.41	6.07	6.00	5.50	5.49	5.19	4.61	5.91
Campeche	4.59	5.45	6.21	6.75	6.92	6.68	6.66	6.54	6.06	5.29	4.75	4.24	5.85
Chetumal	4.06	4.85	5.50	6.04	5.85	5.32	5.34	5.24	4.92	4.60	4.21	3.86	4.98
Chihuahua	4.26	5.02	6.14	6.84	7.26	7.34	6.91	6.39	5.63	5.50	4.16	3.81	5.77
Chilpancingo	5.17	5.98	6.78	6.83	6.23	5.42	5.77	5.61	5.05	5.22	5.18	4.89	5.68
Colima	4.85	5.80	6.92	7.18	6.82	5.73	5.30	5.20	4.85	5.02	5.07	4.61	5.61
Cuernavaca	5.19	6.10	6.96	7.06	6.66	6.01	6.28	6.00	5.43	5.37	5.26	4.90	5.94
Cuidad de México	4.56	5.31	6.00	5.86	5.61	5.47	5.06	5.00	4.53	4.61	4.47	4.22	5.06
Cuidad Victoria	4.02	4.78	5.82	6.03	6.31	6.17	6.11	5.92	5.15	4.82	4.41	3.85	5.28
Culiacán	4.36	5.25	6.55	7.28	7.91	7.68	6.71	6.20	5.68	5.47	4.63	3.99	5.98
Durango	4.42	5.35	6.62	7.01	7.15	6.64	5.97	5.84	5.34	5.40	4.81	4.17	5.73
Guanajuato	4.67	5.64	6.64	6.89	6.85	6.36	6.06	6.01	5.42	5.31	5.05	4.57	5.79
Hermosillo	3.80	4.66	6.19	7.31	7.72	7.71	6.69	6.14	5.81	5.06	4.17	3.54	5.73
La Paz	3.80	4.74	5.96	6.79	7.36	7.30	6.71	6.16	5.55	5.02	4.15	3.54	5.59
Mérida	4.25	4.97	5.77	6.35	6.31	5.87	5.90	5.71	5.36	4.78	4.33	3.98	5.30
Monterrey	3.83	4.61	5.73	5.94	6.27	6.19	6.06	5.74	5.05	4.66	4.20	3.64	5.16
Morelia	2.97	3.89	4.28	4.47	4.36	4.25	4.47	4.22	3.75	3.69	3.56	3.36	3.94
Oaxaca	4.70	5.30	6.11	6.38	6.08	5.33	5.34	5.28	4.70	4.71	4.63	4.53	5.26
Puebla	4.73	5.50	6.20	6.21	6.16	5.64	5.67	5.57	4.95	4.94	4.79	4.49	5.40

Daliy solar irradiation on horizontal surface [kWh/(m2.d) or Peak Sunlight Hours]

City	Jan.	Feb.	Mar.	Apr.	May.	Jun.	Jul.	Aug.	Sep.	Oct.	Nov.	Dec.	Annual Ave.
Querétaro	4.84	5.86	6.81	7.04	6.81	6.36	6.14	6.06	5.49	5.29	5.09	4.58	5.86
San Luis Potosi	4.25	5.11	6.10	6.44	6.66	6.39	6.06	6.03	5.14	5.00	4.62	4.07	5.49
Tepic	4.64	5.63	6.82	7.38	7.66	6.58	5.86	5.76	5.33	5.43	5.06	4.40	5.88
Tuxtla Gutiérrez	4.33	5.01	5.92	6.15	5.90	5.32	5.64	5.45	4.74	4.52	4.50	4.28	5.15
Villahermosa	3.83	4.51	5.47	5.99	5.85	5.49	5.70	5.56	4.85	4.35	4.06	3.61	4.94
Zacatecas	4.57	5.51	6.62	6.95	7.00	6.36	6.02	5.95	5.41	5.34	5.02	4.41	5.76
GUATEMALA													
Cobán	3.92	4.78	6.33	6.11	6.00	5.42	5.67	5.67	5.17	4.08	3.81	3.78	5.06
Gualán	3.79	4.63	5.42	5.82	5.51	5.26	5.16	5.26	4.92	4.23	3.82	3.52	4.78
Guatemala Ciudad	5.18	5.73	6.02	6.05	5.48	5.16	5.45	5.34	4.73	4.76	4.90	4.95	5.31
Huehuetenango	5.86	6.67	7.47	7.39	6.67	7.03	7.58	7.31	6.44	5.89	5.28	5.56	6.60
Puerto Barrios	3.75	4.47	5.50	5.36	5.64	5.56	5.78	5.39	4.81	4.03	3.64	3.89	4.82
Quezaltenango	5.31	5.76	6.07	6.03	5.55	5.31	5.55	5.28	4.70	4.80	5.00	5.12	5.37
San Benito	3.83	4.72	5.52	6.02	5.84	5.41	5.11	5.14	4.94	4.37	3.97	3.58	4.87
San José	6.04	6.70	7.11	7.05	6.24	5.95	6.33	6.30	5.63	5.78	5.80	5.71	6.22
Sayaché	3.95	4.72	5.57	6.03	5.95	5.54	5.48	5.67	5.31	4.60	4.09	3.67	5.05
EL SALVADOR													
Ahuachapan	5.23	5.70	5.94	5.88	5.31	5.17	5.78	5.43	4.75	4.94	5.02	4.99	5.34
La Union	4.96	5.48	5.76	5.56	5.09	5.21	5.62	5.51	4.97	4.90	4.78	4.79	5.22
Nueva Concepción	4.95	5.51	5.76	5.70	5.44	5.49	5.68	5.53	5.20	5.21	4.99	4.74	5.35

Daliy solar irradiation on horizontal surface [kWh/(m2.d) or Peak Sunlight Hours]

City	Jan.	Feb.	Mar.	Apr.	May.	Jun.	Jul.	Aug.	Sep.	Oct.	Nov.	Dec.	Annual Ave.
San Miguel	5.64	6.16	6.50	6.40	5.78	5.75	6.27	6.17	5.29	5.34	5.47	5.39	5.85
San Salvador	4.88	5.30	5.52	5.46	4.73	4.80	5.58	5.36	4.70	4.73	4.81	4.79	5.06
Santa Ana	5.03	5.71	6.16	6.26	5.60	5.42	5.78	5.68	5.07	4.92	4.93	4.79	5.45
HONDURAS													
Brus Laguna	3.98	4.75	5.51	6.03	5.57	4.93	4.56	4.88	5.11	4.47	3.83	3.73	4.78
Catacamas	4.04	4.75	5.54	5.88	5.47	4.96	4.59	4.96	5.09	4.61	4.14	3.84	4.82
Choloma	3.96	4.87	5.68	6.12	5.81	5.74	5.61	5.72	5.45	4.46	3.84	3.60	5.07
Choluteca	5.17	5.69	6.25	6.11	5.97	5.33	6.19	6.03	4.94	5.17	5.14	5.11	5.59
Danli	4.24	5.00	5.85	6.01	5.57	5.39	5.17	5.40	5.26	4.67	4.23	4.00	5.07
Guanaja	4.54	5.60	6.49	7.15	6.80	6.52	6.17	6.36	6.20	5.19	4.39	4.20	5.80
Islas del Cisne	4.52	5.42	6.39	7.02	6.72	6.14	6.57	6.22	5.71	4.77	4.34	4.23	5.67
La Ceiba	3.98	4.87	5.68	6.14	5.70	5.63	5.54	5.65	5.41	4.47	3.85	3.66	5.05
Puerto Lempira	4.57	5.39	6.11	6.63	6.03	5.24	5.18	5.41	5.35	4.73	4.23	4.20	5.26
Roatán	4.49	5.55	6.45	7.09	6.76	6.58	6.44	6.55	6.18	5.18	4.42	4.13	5.82
San Pedro Sula	3.96	4.87	5.68	6.12	5.81	5.74	5.61	5.72	5.45	4.46	3.84	3.60	5.07
Santa Rosa de Copán	4.60	5.39	6.03	6.22	5.65	5.53	5.69	5.58	5.14	4.65	4.55	4.39	5.29
Tegucigalpa	4.42	5.14	5.83	5.81	5.78	5.14	5.56	5.61	5.00	4.86	4.50	4.39	5.17
Trujillo	3.94	4.80	5.65	6.13	5.70	5.45	5.19	5.42	5.38	4.47	3.80	3.60	4.96
NICARAGUA													
Bluefields	4.14	4.89	5.69	5.61	5.39	4.33	4.50	4.61	4.44	4.28	3.94	4.14	4.66

Daliy solar irradiation on horizontal surface [kWh/(m2.d) or Peak Sunlight Hours]

City	Jan.	Feb.	Mar.	Apr.	May.	Jun.	Jul.	Aug.	Sep.	Oct.	Nov.	Dec.	Annual Ave.
Bonanza	4.03	4.68	5.41	5.76	5.31	4.47	4.08	4.38	4.77	4.41	4.00	3.84	4.60
Chinandega	4.72	5.56	5.83	5.69	4.86	4.89	5.17	5.28	4.83	4.86	4.72	4.69	5.09
El Rama	4.28	4.84	5.48	5.66	5.00	4.29	3.84	4.12	4.60	4.41	4.14	4.10	4.56
Jinotega	4.28	4.93	5.73	5.91	5.27	4.84	4.55	4.78	4.92	4.67	4.27	4.09	4.85
Managua A.C. Sandino	5.55	6.06	6.69	6.52	5.77	5.69	5.78	5.75	5.36	5.27	5.24	5.33	5.75
Masaya	6.24	6.84	7.40	7.28	6.37	5.95	6.26	6.27	5.76	5.74	5.74	5.84	6.31
Muy Muy	4.47	5.25	5.14	5.36	4.33	4.81	3.42	4.17	4.97	5.22	4.64	4.14	4.66
Nueva Guinea	4.55	5.05	5.72	5.72	5.03	4.43	4.01	4.25	4.62	4.38	4.23	4.29	4.69
Ocotal	4.47	5.17	5.58	5.56	5.03	5.17	5.14	5.44	5.25	4.69	4.39	4.11	5.00
Puerto Cabezas	4.09	4.73	5.27	5.64	5.13	4.34	4.16	4.27	4.51	4.12	3.93	3.86	4.50
Rivas	5.14	5.81	6.33	6.31	5.56	4.78	5.00	5.03	4.78	4.75	4.47	4.69	5.22
Siuna	4.20	4.73	5.35	5.60	5.09	4.30	3.92	4.24	4.66	4.48	4.20	3.98	4.56
COSTA RICA													
Guácimo	5.86	6.50	6.65	5.67	4.82	4.59	4.64	4.62	4.43	4.19	4.33	5.09	5.12
Liberia	5.11	5.72	6.03	6.06	5.17	4.58	4.94	4.75	4.58	4.42	4.42	4.69	5.04
Paraíso	4.75	5.25	5.78	5.16	4.38	4.35	4.20	4.09	4.27	3.90	3.88	4.33	4.53
Puerto Limon	3.75	4.19	4.50	4.67	4.50	3.92	3.86	4.14	4.31	4.28	3.58	3.31	4.08
Puntarenas	5.92	6.65	7.02	6.42	5.23	4.85	4.87	4.87	4.74	4.56	4.63	5.30	5.42
San Jose / Santamaria	5.11	5.72	6.17	5.81	5.25	4.58	4.89	5.00	4.58	4.53	4.50	4.89	5.09

Daliy solar irradiation on horizontal surface [kWh/(m2.d)] or Peak Sunlight Hours]

City	Jan.	Feb.	Mar.	Apr.	May.	Jun.	Jul.	Aug.	Sep.	Oct.	Nov.	Dec.	Annual Ave.
PANAMÁ													
Changuinola	4.09	4.55	4.87	4.58	4.13	4.04	3.93	3.90	4.12	3.92	3.65	3.74	4.13
Chitrñe	5.53	6.06	6.26	5.68	4.60	4.11	4.12	4.13	4.03	3.87	4.10	4.71	4.77
David	4.94	5.56	5.61	5.03	4.44	4.00	4.19	4.28	4.28	4.00	5.11	4.42	4.66
Jaqué	5.98	6.38	6.66	6.01	4.69	3.96	4.19	4.23	4.16	4.12	4.24	4.87	4.96
Kusapin	4.52	5.15	5.58	5.32	4.89	4.25	3.88	4.12	4.75	4.41	3.88	3.83	4.55
La Palma	5.56	5.98	6.32	5.81	4.83	4.27	4.36	4.43	4.35	4.28	4.30	4.73	4.94
Marcos A Gelabert I	5.09	5.70	6.01	5.68	4.63	4.04	4.15	4.17	4.23	4.19	4.21	4.35	4.70
Puerto Escondido	5.05	5.66	6.02	5.62	4.77	4.15	3.98	4.23	4.60	4.29	3.82	4.10	4.69
Río de Jesús	6.11	6.64	6.65	5.96	4.86	4.52	4.66	4.57	4.29	4.23	4.51	5.20	5.18
Santiago	4.83	5.44	6.00	5.28	4.86	4.39	4.69	4.75	4.39	4.31	4.42	4.42	4.82
Soná	4.54	5.00	5.44	5.02	4.20	4.04	3.97	3.84	4.00	3.77	3.74	4.06	4.30
Tocumen	4.97	5.39	5.22	5.39	4.64	4.42	4.50	4.33	4.72	4.47	4.33	5.06	4.79
Ukupseni	5.18	5.46	5.82	5.45	4.57	4.21	4.33	4.45	4.54	4.33	4.06	4.37	4.73
Ustupo	5.53	5.73	6.00	5.74	4.99	4.70	4.90	4.94	4.85	4.63	4.42	4.72	5.10
COLOMBIA													
Arauca	5.24	5.35	5.19	4.70	4.40	4.25	4.45	4.70	5.05	4.99	4.79	4.87	4.83
Baranquilla	5.63	5.70	5.89	5.51	5.08	5.24	5.42	5.36	4.97	4.68	4.72	5.04	5.27
Bogota	5.01	4.66	4.47	3.93	3.70	3.79	4.04	4.33	4.31	4.33	4.10	4.55	4.27
Cali	4.50	4.61	4.83	4.61	4.58	4.42	5.14	5.14	4.69	4.53	4.39	4.42	4.66

Daliy solar irradiation on horizontal surface [kWh/(m2.d) or Peak Sunlight Hours]

City	Jan.	Feb.	Mar.	Apr.	May.	Jun.	Jul.	Aug.	Sep.	Oct.	Nov.	Dec.	Annual Ave.
Cartagena	5.08	4.63	4.30	4.00	4.03	3.75	3.82	4.22	4.70	4.70	4.72	4.94	4.41
Florencia	4.24	4.02	3.78	3.78	3.79	3.57	3.55	3.73	4.18	4.30	4.21	4.18	3.94
Leticia	4.44	4.61	4.67	4.44	4.47	4.39	4.83	5.11	5.06	5.11	4.89	4.61	4.72
Manizales	4.35	4.51	4.49	4.31	4.44	4.52	4.93	4.95	4.70	4.47	4.30	4.12	4.51
Medellín	4.33	4.61	4.72	4.64	4.58	4.81	5.56	5.22	4.69	4.39	4.33	4.28	4.68
Mitú	5.18	4.83	4.76	4.42	4.18	3.92	4.00	4.46	4.81	4.76	4.64	4.81	4.56
Montería	4.89	4.95	4.97	4.70	4.61	4.75	5.03	4.98	4.76	4.64	4.50	4.54	4.78
Neiva	4.69	4.72	4.75	4.58	4.78	4.58	5.00	4.86	4.81	4.86	4.69	4.72	4.75
Pasto	4.06	4.25	4.44	4.28	4.06	4.03	4.22	4.21	4.11	4.10	3.90	3.84	4.13
Pereira	4.55	4.72	4.77	4.52	4.52	4.60	4.92	4.97	4.80	4.56	4.44	4.37	4.65
Popayán	4.42	4.54	4.64	4.44	4.37	4.46	4.66	4.70	4.60	4.41	4.31	4.27	4.49
Puerto Carreño	5.96	6.11	5.95	5.24	4.66	4.35	4.52	4.79	5.06	5.20	5.24	5.52	5.22
Quibdó	3.72	3.98	4.23	3.80	3.93	3.91	4.34	4.29	4.06	3.86	3.73	3.48	3.94
Rio Sucio	4.63	4.90	5.06	4.43	3.97	3.78	4.13	4.01	3.86	3.79	3.77	4.02	4.20
San Andres	5.35	6.12	6.90	7.01	6.07	5.37	5.48	5.56	5.39	4.94	4.59	4.74	5.63
San José del Guavire	5.37	5.06	4.83	4.33	4.18	3.95	4.11	4.58	4.95	4.68	4.60	4.91	4.63
Santa Marta	6.17	6.63	6.99	6.84	6.23	6.29	6.62	6.56	6.04	5.59	5.44	5.59	6.25
Sincelejo	5.79	5.90	5.97	5.63	5.09	5.21	5.44	5.32	4.94	4.67	4.75	5.14	5.32
Tunja	5.18	5.15	5.11	4.78	4.84	5.02	5.17	5.22	5.31	4.91	4.82	4.91	5.04
Valledupar	5.46	5.53	5.74	5.39	5.17	5.29	5.47	5.28	5.08	4.77	4.82	5.00	5.25

Daliy solar irradiation on horizontal surface [kWh/(m2.d)] or Peak Sunlight Hours]

City	Jan.	Feb.	Mar.	Apr.	May.	Jun.	Jul.	Aug.	Sep.	Oct.	Nov.	Dec.	Annual Ave.
Villavicencio	4.72	4.56	4.40	4.06	4.16	4.16	4.09	4.15	4.50	4.33	4.27	4.43	4.32
VENEZUELA													
Barcelona	4.60	5.06	5.55	5.36	5.14	4.71	4.74	4.89	4.96	4.66	4.29	4.29	4.85
Barinas	4.66	4.70	4.66	4.31	4.27	4.37	4.51	4.56	4.71	4.50	4.29	4.36	4.49
Barquisimeto	4.96	5.26	5.39	4.89	4.95	5.15	5.33	5.52	5.36	4.97	4.76	4.63	5.10
Caracas	4.82	5.36	5.83	5.51	5.40	5.51	5.58	5.67	5.59	4.86	4.38	4.38	5.24
Ciudad Bolívar	4.39	4.75	5.10	5.02	4.81	4.49	4.75	4.96	4.99	4.69	4.29	4.14	4.70
Coro	5.25	5.78	6.27	6.02	5.92	5.86	5.95	6.22	5.99	5.36	4.97	4.77	5.70
Guaira	4.42	5.08	5.54	5.72	5.44	4.85	5.16	5.31	5.42	4.96	4.30	4.03	5.02
Guanare	4.83	5.11	4.92	4.50	4.53	4.61	4.94	5.22	5.28	5.17	4.69	4.61	4.87
La Asunción	5.76	6.46	7.10	7.30	7.00	6.64	6.74	6.90	6.85	6.33	5.66	5.43	6.51
Maracaibo	4.02	4.32	4.59	4.21	4.10	4.30	4.62	4.54	4.27	3.87	3.63	3.66	4.18
Maracay	4.65	5.09	5.41	5.01	4.77	4.67	4.78	4.92	5.02	4.79	4.49	4.40	4.83
Maturín	4.10	4.40	4.74	4.62	4.46	4.04	4.19	4.49	4.63	4.39	3.96	3.87	4.32
Mérida	5.08	5.31	5.37	4.99	4.96	4.71	4.97	5.25	5.25	4.83	4.62	4.74	5.01
Puerto Ayacucho	4.36	4.61	4.61	4.14	3.82	3.66	3.84	3.93	4.19	4.23	4.15	4.14	4.14
San Carlos	5.15	5.50	5.79	5.50	5.35	5.46	5.72	5.89	5.61	5.16	4.78	4.73	5.39
San Fernando	5.04	5.44	5.80	5.34	5.00	4.54	4.62	4.91	5.26	5.14	4.92	4.81	5.07
San Juan de los Morros	5.39	5.75	6.04	5.57	5.11	4.93	5.02	5.18	5.34	5.27	5.09	5.07	5.31
Tucupita	4.75	5.10	5.59	5.65	5.14	4.72	4.86	5.12	5.20	4.98	4.51	4.39	5.00

Daliy solar irradiation on horizontal surface [kWh/(m2.d) or Peak Sunlight Hours]

City	Jan.	Feb.	Mar.	Apr.	May.	Jun.	Jul.	Aug.	Sep.	Oct.	Nov.	Dec.	Annual Ave.
ECUADOR													
Ambato	4.39	4.46	4.64	4.45	4.26	4.23	4.32	4.64	4.62	4.61	4.70	4.45	4.48
Azogues	4.39	4.25	4.45	4.33	4.19	4.15	4.19	4.49	4.55	4.54	4.73	4.57	4.40
Esmeraldas	4.24	4.43	5.00	4.80	4.22	3.73	3.87	4.06	4.21	3.98	3.81	4.07	4.20
Guayaquil	3.42	4.42	3.39	4.36	4.33	3.58	4.36	3.64	5.69	4.17	3.72	4.61	4.14
Ibarra	3.96	4.09	4.35	4.17	3.91	3.80	4.03	4.05	3.88	3.86	3.74	3.69	3.96
Machala	4.83	4.80	5.19	4.88	4.65	4.60	4.65	4.93	5.18	4.95	5.26	5.07	4.92
Puyo	3.68	3.48	3.44	3.63	3.62	3.48	3.53	3.81	4.07	4.22	4.16	3.83	3.75
Quito	4.14	4.35	4.55	4.33	4.12	4.02	4.27	4.46	4.27	4.24	4.30	3.98	4.25
Santo Domingo	3.96	4.30	4.73	4.50	3.93	3.57	3.62	4.02	4.10	3.81	3.76	3.78	4.01
Tulcán	3.86	3.66	3.64	3.71	3.75	3.76	3.82	4.04	4.11	4.11	4.15	3.95	3.88
Zamora	3.99	3.81	4.05	3.98	4.05	4.02	3.98	4.29	4.62	4.68	4.80	4.40	4.22
PERÚ													
Abancay	5.51	5.52	5.34	5.20	5.29	5.07	5.18	5.37	5.72	6.00	6.36	5.94	5.54
Barranca	6.61	6.58	6.73	6.18	5.15	4.09	4.17	4.57	5.36	6.08	6.61	6.84	5.75
Cajamarca	5.54	5.30	5.62	5.40	5.28	5.16	5.33	5.72	6.04	6.18	6.35	6.00	5.66
Camaná	6.51	6.38	6.07	5.24	4.33	3.79	3.82	4.38	5.17	6.06	6.63	6.83	5.43
Celendín	4.82	4.61	4.88	4.71	4.93	4.93	5.06	5.46	5.76	5.67	5.86	5.37	5.17
Cerro de Pasco	5.43	5.40	5.35	5.37	5.48	5.34	5.56	5.95	6.16	6.14	6.40	5.95	5.71
Chepén	6.60	6.63	6.72	6.40	5.40	4.23	4.11	4.31	4.95	5.78	6.12	6.39	5.64

Daliy solar irradiation on horizontal surface [kWh/(m2.d) or Peak Sunlight Hours]

City	Jan.	Feb.	Mar.	Apr.	May.	Jun.	Jul.	Aug.	Sep.	Oct.	Nov.	Dec.	Annual Ave.
Chiclayo	5.70	5.55	5.84	5.38	4.89	4.47	4.48	4.91	5.59	5.95	5.98	5.86	5.38
Chimbote	7.07	7.09	6.98	6.34	5.12	3.72	3.67	3.94	4.42	5.42	6.22	6.83	5.57
Chivay	5.85	5.70	5.56	5.64	5.58	5.27	5.45	5.90	6.58	6.96	7.26	6.65	6.03
Contamana	4.58	4.26	4.14	4.28	4.39	4.34	4.65	4.95	5.00	4.96	4.82	4.74	4.59
Coracora	6.35	6.08	6.06	5.77	5.31	4.90	5.03	5.54	6.34	6.99	7.29	7.00	6.06
Cuzco	5.19	4.69	4.92	5.03	4.81	4.78	5.06	5.11	5.17	5.64	5.75	5.22	5.11
Huancavelica	5.14	5.16	4.91	5.01	5.21	5.07	5.19	5.44	5.58	5.74	5.91	5.55	5.33
Huancayo	7.38	6.71	6.54	6.54	6.17	6.28	6.30	6.74	7.17	7.47	7.75	7.21	6.86
Huánuco	4.95	4.83	4.76	4.76	4.97	5.03	5.14	5.35	5.29	5.27	5.52	5.28	5.10
Huaraz	5.98	5.88	6.09	5.75	5.45	5.03	5.25	5.76	6.23	6.41	6.81	6.52	5.93
Huarmey	7.26	7.22	7.06	6.29	4.82	3.32	3.32	3.64	4.23	5.08	5.90	6.76	5.41
Ica	6.68	6.70	6.64	5.92	5.12	4.32	4.29	4.78	5.79	6.59	6.92	7.00	5.90
Llave	6.00	6.06	5.73	5.72	5.60	5.28	5.43	5.86	6.59	6.99	7.18	6.68	6.09
Llo	7.06	7.15	6.49	4.85	3.67	2.93	2.89	3.32	4.19	5.25	6.22	6.84	5.07
Iquitos	4.66	4.48	4.51	4.31	4.19	4.06	4.35	4.76	5.04	4.92	4.79	4.63	4.56
Jaén	4.16	3.93	4.25	4.24	4.31	4.22	4.31	4.62	4.95	4.94	5.15	4.63	4.48
Juanjí	4.33	4.05	3.96	4.11	4.25	4.21	4.40	4.64	4.60	4.74	4.70	4.49	4.37
Juliaca	5.84	5.84	5.66	5.65	5.63	5.39	5.57	5.94	6.49	6.84	7.01	6.50	6.03
Lima	7.14	7.15	7.04	6.33	4.93	3.39	3.24	3.58	4.32	5.29	6.01	6.80	5.44
Mollendo	6.86	6.82	6.01	4.60	3.50	2.84	2.81	3.22	4.25	5.22	6.21	6.71	4.92

Daliy solar irradiation on horizontal surface [kWh/(m2.d) or Peak Sunlight Hours]

City	Jan.	Feb.	Mar.	Apr.	May.	Jun.	Jul.	Aug.	Sep.	Oct.	Nov.	Dec.	Annual Ave.
Moquegua	6.85	6.68	6.29	5.39	4.63	4.01	4.21	4.78	5.53	6.50	6.99	7.13	5.75
Nauta	4.60	4.47	4.49	4.30	4.27	4.14	4.46	4.84	5.02	4.98	4.80	4.57	4.58
Nazca	6.20	5.97	6.01	5.83	5.45	5.06	5.28	5.80	6.63	7.14	7.27	6.86	6.13
Oxapampa	4.53	4.48	4.46	4.51	4.71	4.74	4.80	4.92	4.90	4.99	5.16	4.88	4.76
Paita	6.96	6.97	6.83	6.64	5.66	4.62	4.39	4.77	5.34	5.90	6.23	6.66	5.91
Pisco	7.36	7.44	7.23	6.50	5.30	3.97	3.63	4.01	5.21	6.45	6.93	7.23	5.94
Piura	6.11	6.11	6.26	5.96	5.32	4.76	4.63	4.99	5.68	5.90	6.02	6.19	5.66
Pucallpa	4.54	4.32	4.20	4.31	4.31	4.23	4.71	5.13	5.22	5.09	4.77	4.68	4.63
Requena	4.64	4.47	4.45	4.26	4.30	4.24	4.56	4.94	5.14	5.08	4.78	4.64	4.63
Rioja	4.45	4.23	4.32	4.31	4.38	4.38	4.42	4.60	4.78	4.76	5.04	4.75	4.54
Satipo	4.12	4.25	4.13	4.25	4.44	4.36	4.48	4.61	4.56	4.67	4.72	4.47	4.42
Sicuani	5.84	5.76	5.58	5.52	5.64	5.35	5.57	5.85	6.26	6.56	6.77	6.35	5.92
Sullana	5.54	5.31	5.66	5.41	5.13	5.15	5.15	5.62	6.01	6.15	6.16	5.91	5.60
Tacna	7.45	7.40	6.59	4.87	3.61	2.83	2.92	3.35	4.38	5.49	6.43	7.19	5.21
Talara	6.85	6.81	6.79	6.68	5.95	5.21	5.03	5.27	5.85	6.29	6.61	6.75	6.17
Tarapoto	4.75	4.67	4.36	4.28	4.17	4.06	4.58	4.83	4.72	4.72	4.89	4.83	4.57
Tarma	4.97	4.97	4.79	4.91	5.13	5.14	5.24	5.52	5.60	5.59	5.82	5.37	5.25
Tocache	4.44	4.29	4.26	4.35	4.53	4.56	4.66	4.82	4.67	4.74	4.90	4.74	4.58
Trujillo	7.13	7.05	6.95	6.40	5.27	3.88	3.79	4.02	4.45	5.27	6.06	6.70	5.58
Tumbes	6.23	6.16	6.32	6.38	5.97	5.56	5.31	5.54	6.10	6.18	6.46	6.40	6.05

Daily solar irradiation on horizontal surface [kWh/(m2.d) or Peak Sunlight Hours]

City	Jan.	Feb.	Mar.	Apr.	May.	Jun.	Jul.	Aug.	Sep.	Oct.	Nov.	Dec.	Annual Ave.
Virú	6.39	6.40	6.59	6.17	5.35	4.48	4.47	4.74	5.27	5.99	6.43	6.55	5.74
Yurimaguas	3.86	3.81	3.70	3.95	4.03	3.95	4.32	4.67	4.74	4.72	4.45	4.04	4.19
BOLIVIA													
Aiquile	5.87	6.04	5.73	5.64	5.41	5.03	5.30	5.73	6.30	6.60	6.80	6.39	5.90
Apolo	4.17	4.47	5.72	4.00	3.89	3.50	4.31	4.44	4.58	5.42	5.00	5.06	4.55
Ascención	4.99	4.94	4.81	4.68	4.11	4.10	4.57	4.77	5.15	5.46	5.40	5.14	4.84
Camargo	6.59	6.53	6.15	5.87	5.34	4.90	5.20	5.83	6.72	7.25	7.46	7.14	6.25
Caranavi	4.68	4.77	4.57	4.36	4.07	3.92	3.95	4.41	4.69	5.07	5.05	4.93	4.54
Chulumani	5.31	5.47	5.30	5.01	4.95	4.65	4.71	5.01	5.55	5.78	6.04	5.78	5.30
Cobija	4.75	4.63	4.49	4.56	4.26	4.37	4.94	5.29	5.24	5.28	4.98	4.80	4.80
Cochabamba	5.25	5.92	5.22	5.22	5.06	4.56	4.94	5.50	5.92	6.39	6.67	5.83	5.54
La Paz	4.86	5.22	4.78	5.08	5.31	4.67	4.94	5.44	5.64	5.94	6.28	5.33	5.29
Magdalena	5.00	4.95	4.84	4.79	4.34	4.46	4.86	5.07	5.19	5.45	5.34	5.15	4.95
Monteaguado	5.65	5.41	5.03	4.51	4.11	4.16	4.37	4.92	5.49	5.56	5.74	5.63	5.05
Oruro	6.31	6.42	6.06	5.80	5.56	5.24	5.41	5.82	6.47	6.93	7.16	6.89	6.17
Puerto Suárez	5.64	5.40	5.09	4.74	4.00	3.91	4.36	4.77	4.97	5.41	5.73	5.73	4.98
Punata	5.34	5.39	5.38	5.22	5.05	4.75	4.82	5.15	5.64	5.93	6.16	5.77	5.38
Riberalta	4.58	4.60	4.53	4.52	4.31	4.48	5.04	5.28	5.21	5.27	5.01	4.71	4.80
San Ignacio	5.05	4.99	4.85	4.77	4.15	4.08	4.54	4.97	5.31	5.59	5.46	5.19	4.91
San Javier	5.24	5.15	4.98	4.58	3.95	3.90	4.37	4.72	5.10	5.43	5.50	5.15	4.84

Daily solar irradiation on horizontal surface [kWh/(m2.d) or Peak Sunlight Hours]

City	Jan.	Feb.	Mar.	Apr.	May.	Jun.	Jul.	Aug.	Sep.	Oct.	Nov.	Dec.	Annual Ave.
San José	5.59	5.33	5.08	4.62	3.85	3.76	4.25	4.63	4.94	5.35	5.58	5.56	4.88
San Matías	5.41	5.25	5.03	4.83	4.17	4.20	4.67	4.97	5.03	5.47	5.61	5.46	5.01
Santa Ana	5.03	4.91	4.83	4.72	4.27	4.31	4.76	5.07	5.23	5.46	5.28	5.10	4.91
Santa Cruz	5.50	5.19	5.03	4.19	3.53	3.03	3.69	4.50	4.53	5.56	5.92	5.61	4.69
Santa Rosa	5.23	5.06	4.87	4.51	3.92	3.93	4.41	4.70	4.99	5.29	5.34	5.13	4.78
Sucre	5.42	5.22	5.06	4.72	5.00	4.47	4.81	5.42	5.97	6.36	6.50	5.67	5.38
Tarija	6.27	6.06	5.82	5.50	4.98	4.71	4.86	5.51	6.44	6.60	6.83	6.71	5.86
Trinidad	4.78	4.64	4.28	4.33	3.64	4.00	4.25	4.72	4.00	5.33	5.97	4.94	4.57
Tupiza	6.79	6.70	6.33	6.02	5.41	4.94	5.20	5.87	6.93	7.42	7.63	7.29	6.38
Uyuni	6.66	6.55	5.99	5.26	4.71	4.18	4.45	4.93	5.89	6.78	7.14	7.07	5.80
Vallegrande	5.92	5.92	5.65	5.47	5.20	4.94	5.09	5.60	6.21	6.45	6.64	6.33	5.79
Villa Serrano	5.88	5.81	5.58	5.44	5.13	4.86	5.05	5.55	6.25	6.31	6.51	6.27	5.72
Yacuíba	5.72	5.30	4.50	3.64	3.30	3.17	3.64	4.40	5.00	5.32	5.50	5.69	4.60
CHILE													
Antofagasta	6.84	6.62	5.73	4.60	3.66	3.17	3.40	4.05	5.05	6.08	6.74	7.03	5.25
Arica	6.24	6.21	5.63	4.55	3.66	2.95	2.92	3.39	4.30	5.36	6.08	6.29	4.80
Cayhaique	4.46	4.40	3.11	1.97	1.28	0.97	1.12	1.68	2.63	3.64	4.16	4.49	2.83
Concepción	6.96	5.89	4.63	3.05	1.78	1.44	1.68	2.47	3.66	5.11	6.22	6.79	4.14
Copiapó	7.23	6.69	5.58	4.21	3.36	2.98	3.31	4.10	5.25	6.39	7.07	7.33	5.29
Iquique	6.53	6.33	5.44	4.53	3.42	2.83	2.72	3.25	4.03	5.19	6.08	6.69	4.75

Daliy solar irradiation on horizontal surface [kWh/(m2.d) or Peak Sunlight Hours]

City	Jan.	Feb.	Mar.	Apr.	May.	Jun.	Jul.	Aug.	Sep.	Oct.	Nov.	Dec.	Annual Ave.
La Serena	6.22	5.88	4.46	3.14	2.49	2.33	2.48	2.92	4.05	4.78	5.48	6.33	4.21
Puerto Montt	5.59	5.31	3.51	2.36	1.39	1.01	1.21	1.92	2.79	3.72	4.70	5.33	3.24
Punta Arenas	5.50	4.35	2.96	1.65	0.83	0.55	0.69	1.39	2.57	4.25	5.44	5.75	2.99
Rancagua	8.74	7.66	6.03	4.24	2.78	2.22	2.50	3.34	4.70	6.19	7.82	8.76	5.42
Santiago	6.53	5.72	4.38	2.80	1.74	1.28	1.41	2.32	3.34	4.85	5.87	6.70	3.91
Talca	8.34	7.23	5.65	3.94	2.53	2.01	2.29	3.12	4.58	6.04	7.59	8.39	5.14
Temuco	6.89	5.83	4.46	2.71	1.54	1.23	1.33	2.16	3.31	4.85	5.84	6.81	3.91
Valdivia	7.68	6.73	5.08	3.37	2.15	1.69	1.98	2.82	4.18	5.68	6.83	7.62	4.65
Valparaíso	6.10	4.99	4.03	2.82	1.84	1.42	1.79	2.50	3.55	4.43	5.54	6.09	3.76
PARAGUAY													
Asunción	6.14	5.86	5.33	4.39	3.61	3.00	3.47	3.86	4.56	5.56	6.00	6.42	4.85
Ayolas	6.64	5.97	5.07	3.91	3.36	2.71	3.15	3.80	4.57	5.44	6.35	6.74	4.81
Concepción	5.67	5.67	5.19	4.19	3.67	3.00	3.53	3.75	4.36	5.36	6.08	6.00	4.71
Encarnación	5.94	5.83	4.86	4.03	3.39	2.72	3.17	3.67	4.39	5.42	6.11	6.44	4.66
Mayor Pablo Lagerenza	6.07	5.70	5.22	4.43	3.58	3.31	3.90	4.45	5.03	5.62	5.88	5.92	4.93
Pedro Juan Caballero	6.03	5.63	5.25	4.58	3.68	3.46	3.83	4.44	4.91	5.50	6.14	6.26	4.98
San Estanislao	6.56	5.92	5.39	4.38	3.54	3.08	3.39	4.06	4.75	5.52	6.31	6.61	4.96
San Pedro	6.62	5.95	5.33	4.31	3.53	3.03	3.40	4.08	4.81	5.54	6.36	6.60	4.96
Santa Rosa	6.62	5.95	5.13	4.04	3.45	2.85	3.19	3.88	4.61	5.43	6.31	6.69	4.85
Villarica	6.28	5.83	5.11	4.44	3.67	3.06	3.33	3.89	4.61	5.56	6.36	6.31	4.87

Daliy solar irradiation on horizontal surface [kWh/(m2.d) or Peak Sunlight Hours]

City	Jan.	Feb.	Mar.	Apr.	May.	Jun.	Jul.	Aug.	Sep.	Oct.	Nov.	Dec.	Annual Ave.
URUGUAY													
Artigas	5.75	4.36	4.22	3.31	2.58	2.25	2.28	2.86	3.69	4.33	5.06	5.53	3.85
Melo	6.71	5.68	4.88	3.65	2.86	2.32	2.58	3.30	4.33	5.26	6.35	6.91	4.57
Colonia del Sacramento	6.92	6.00	4.91	3.60	2.74	2.19	2.46	3.30	4.50	5.38	6.50	7.01	4.63
Durazno	6.94	5.94	4.97	3.62	2.81	2.27	2.53	3.30	4.41	5.33	6.48	6.99	4.63
Rivera	6.74	5.76	4.95	3.74	3.08	2.52	2.81	3.61	4.49	5.35	6.44	6.86	4.70
Rocha	6.73	5.66	4.72	3.47	2.63	2.13	2.33	3.09	4.19	5.09	6.16	6.77	4.41
Salto	6.90	5.94	5.06	3.70	3.00	2.46	2.78	3.59	4.63	5.46	6.52	6.93	4.75
Tacuarembó	6.71	5.72	4.89	3.68	2.96	2.43	2.72	3.46	4.44	5.36	6.43	6.88	4.64
Treinta y Tres	5.72	4.78	4.56	3.08	2.22	1.92	2.31	2.86	3.72	4.75	5.08	5.61	3.88
ARGENTINA													
Buenos Aires	6.91	6.13	4.93	3.65	2.65	2.08	2.25	3.06	4.19	5.16	6.56	6.81	4.53
Corrientes	6.64	5.92	5.00	3.89	3.46	2.74	3.16	3.90	4.81	5.55	6.35	6.70	4.84
Córdoba	7.21	6.34	5.21	4.07	3.13	2.81	3.15	4.07	5.31	6.22	7.00	7.41	5.16
Formosa	6.50	5.83	5.08	4.33	3.61	3.03	3.44	3.89	4.44	5.64	6.58	6.78	4.93
La Rioja	6.70	6.08	5.18	4.38	3.52	3.14	3.43	4.35	5.50	6.37	6.93	7.09	5.22
Mendoza	7.40	6.59	5.60	4.26	3.06	2.44	2.68	3.53	4.60	6.12	7.12	7.52	5.08
Neuquén	6.33	5.89	4.58	3.36	2.33	1.78	2.00	2.94	3.72	5.28	6.33	6.36	4.24
Paraná	6.86	6.06	5.04	3.78	3.01	2.47	2.82	3.72	4.86	5.64	6.60	6.95	4.82
Posadas	6.42	6.03	5.08	4.17	3.19	2.61	2.89	3.53	4.11	5.36	6.36	6.75	4.71

Daliy solar irradiation on horizontal surface [kWh/(m2.d) or Peak Sunlight Hours]

City	Jan.	Feb.	Mar.	Apr.	May.	Jun.	Jul.	Aug.	Sep.	Oct.	Nov.	Dec.	Annual Ave.
Resistencia	6.70	5.96	5.09	3.86	3.43	2.74	3.17	4.00	4.90	5.65	6.39	6.79	4.89
Río Gallegos	5.54	4.71	3.44	2.06	1.15	0.76	0.89	1.66	2.97	4.47	5.51	5.75	3.24
Salta	5.39	4.97	4.08	3.39	3.00	2.75	3.06	3.75	4.25	5.28	5.25	5.44	4.22
San Juan	7.28	6.69	5.64	4.39	3.50	2.78	3.00	3.86	5.06	6.25	7.44	7.69	5.30
San Luis	6.56	6.25	5.17	4.06	3.08	2.56	2.89	3.89	4.50	6.00	6.72	6.92	4.88
Santa Rosa	6.89	6.00	4.75	3.53	2.44	1.97	2.19	2.94	3.86	5.22	6.67	7.03	4.46
Santiago del Estero	6.19	5.69	4.86	3.92	2.97	2.61	3.03	3.83	4.64	5.56	6.22	6.31	4.65
Ushuaia	5.05	3.87	2.58	1.48	0.70	0.40	0.53	1.18	2.45	3.82	4.89	5.39	2.70
Viedma	7.53	6.52	4.95	3.41	2.22	1.73	2.02	2.88	4.18	5.61	7.06	7.63	4.65
CUBA													
Camagüey	3.44	4.17	5.19	5.39	5.00	5.00	5.50	5.28	4.47	3.94	3.44	3.28	4.51
Ciego de Ávila	4.18	4.97	5.69	6.28	6.12	6.07	6.24	5.86	5.31	4.76	4.25	3.96	5.31
Cienfuegos	3.98	4.76	5.47	6.18	5.93	5.61	5.94	5.66	5.00	4.54	3.99	3.70	5.06
Guantánamo	4.42	5.19	6.08	6.67	6.42	6.17	6.58	6.31	5.75	5.00	4.50	4.14	5.60
Holguín	4.17	4.86	5.48	6.04	5.84	5.98	6.05	5.76	5.35	4.77	4.27	3.91	5.21
La Habana	4.04	4.88	5.65	6.41	6.26	5.74	6.14	5.87	5.07	4.77	4.20	3.77	5.23
Pinar del Río	4.01	4.73	5.54	6.26	6.15	5.84	5.85	5.60	5.07	4.67	4.11	3.72	5.13
Sancti Spíritus	4.48	5.39	6.24	6.96	6.69	6.54	6.84	6.38	5.67	5.10	4.54	4.14	5.75
Santiago de Cuba	5.03	5.83	6.54	7.12	6.69	6.79	6.98	6.66	5.99	5.51	4.92	4.61	6.06
Varadero	4.11	5.05	6.03	7.03	7.14	6.78	7.08	6.81	5.91	5.07	4.18	3.73	5.74

Daliy solar irradiation on horizontal surface [kWh/(m2.d) or Peak Sunlight Hours]

City	Jan.	Feb.	Mar.	Apr.	May.	Jun.	Jul.	Aug.	Sep.	Oct.	Nov.	Dec.	Annual Ave.
REPÚBLICA DOMINICANA													
Barahona	4.67	5.28	5.83	6.44	6.11	6.11	5.86	6.00	5.78	5.28	5.00	4.58	5.58
Caucedo de las Ámericas	4.20	4.80	5.48	5.80	5.68	5.67	5.62	5.42	5.10	4.80	4.27	4.01	5.07
La Romana	4.29	5.05	5.81	6.10	6.02	5.97	6.03	5.89	5.23	4.90	4.43	4.17	5.32
Mao	4.44	5.11	5.72	5.96	5.92	6.40	6.34	6.12	5.82	5.19	4.48	4.15	5.47
Nagua	4.29	4.98	5.77	6.12	6.12	6.40	6.37	6.17	5.81	5.22	4.26	4.00	5.46
San Cristóbal	4.35	4.94	5.33	5.57	5.53	5.85	5.94	5.68	5.24	4.78	4.39	4.20	5.15
Santiago	4.17	4.83	5.56	5.92	5.86	6.28	6.22	6.11	5.50	5.11	4.58	4.08	5.35
Santo Domingo	4.20	4.80	5.48	5.80	5.68	5.67	5.62	5.42	5.10	4.80	4.27	4.01	5.07
PUERTO RICO													
Aquadilla	4.12	4.79	5.51	5.73	5.51	5.79	5.80	5.66	5.21	4.76	4.06	3.89	5.07
Eugenio Maria de Ho	3.95	4.36	5.03	5.19	5.10	5.22	5.09	5.03	4.70	4.44	3.85	3.80	4.65
Guayama	5.07	5.88	6.82	7.20	7.03	6.98	6.88	6.86	6.26	5.74	5.01	4.84	6.21
Mayaguez	4.91	5.74	6.67	6.94	6.82	6.88	6.81	6.86	6.25	5.69	4.94	4.67	6.10
Mercedita	3.95	4.42	5.13	5.36	5.13	5.33	5.42	5.29	4.79	4.29	3.83	3.66	4.72
Ponce	4.60	5.29	6.06	6.33	6.23	6.44	6.38	6.19	5.68	5.26	4.55	4.40	5.62
Roosevelt Roads	3.90	4.30	5.13	5.21	5.05	5.13	5.27	5.16	4.90	4.28	3.64	3.58	4.63
San Juan	4.28	4.91	5.72	6.10	5.78	6.05	6.09	5.96	5.53	4.92	4.31	3.97	5.30

ABOUT THE AUTHOR

Oliver Style worked for 10 years in Mexico, Colombia and Ecuador, delivering projects and training programmes in off-grid photovoltaic systems, efficient biomass cooking stoves, and water supply systems, for the NGO Concern America.

Based in Barcelona, Spain, he currently works as a consultant and fund-raiser for Concern America's appropriate technology programmes, alongside work as a Passive House designer.

www.ingramcontent.com/pod-product-compliance
Lightning Source LLC
Chambersburg PA
CBHW050107210326
41519CB00015BA/3853